Marysol Dantas de Medeiros

Social and environmental vulnerability in the municipality of Natal/RN

Marysol Dantas de Medeiros

Social and environmental vulnerability in the municipality of Natal/RN

Social vulnerability and exposure to natural risks

ScienciaScripts

Imprint

Cover image: www.ingimage.com

This book is a translation from the original published under ISBN 978-3-330-75835-3.

Publisher:
Sciencia Scripts
is a trademark of
Dodo Books Indian Ocean Ltd. and OmniScriptum S.R.L publishing group

120 High Road, East Finchley, London, N2 9ED, United Kingdom
Str. Armeneasca 28/1, office 1, Chisinau MD-2012, Republic of Moldova, Europe
Printed at: see last page
ISBN: 978-620-8-29697-1

CONTENTS

1 INTRODUCTION

Today's society is surrounded by various types of risk, causing individuals to feel a constant sense of fear and insecurity, since the negative results of these risks always bring some kind of damage to the population directly or indirectly involved.

Studies by the United Nations (UN) show that half of the world's population is in urban areas, a third of which lives in slums and informal settlements, especially in developing countries, where migration from the countryside to urban centers has occurred throughout history, in most cases without the necessary time to adapt urban areas in order to absorb this contingent in an orderly and safe manner.

This can be seen in Brazil, which is not prone to catastrophic natural events of great magnitude, such as strong earthquakes, volcanisms and tsunamis. However, it is considered to be one of the countries with the highest number of victims of natural disasters in the world, mainly due to climatic phenomena, but its consequences, such as floods and landslides, are in most cases the result of human action, which degrades and occupies risk areas.

Unequal access to and acquisition of urban land due to the conditions imposed by capitalism end up leading the poorest populations to seek out the areas that have been neglected by real estate speculation, spatially reflecting a pattern of urban occupation, with socially segregated areas being visibly distinguished.

Not only must the location of the dwelling be taken into account, but this factor, coupled with poverty, means that individuals in this situation are more likely to have few basic infrastructures necessary for their safety, such as dwellings built with suitable materials; generally these areas do not have public services such as sanitary sewage, water supply, garbage collection, which further aggravates the formation of "risk zones"[1] .

The capital of Rio Grande do Sul is no exception to this rule, as for several decades it has "housed" the city's poorest population in places unsuitable for construction, such as steep slopes, dune fields, river banks and mangrove areas, which are environments with strong restrictions on use .[2]

1 Almeida, 2010.

2 These environments are classified by the New Brazilian Forest Code (Law No. 12.651, of May 25, 2012) as Permanent Preservation Areas (APP), which have the function of preserving water resources, the landscape,

Although there are no significant records of natural disasters, the outskirts of the city, where housing is notoriously unhealthy and where environmental degradation is constant, suffer from a variety of problems caused by natural hazards, which culminate, in most cases, in financial losses, especially for the families affected, thus characterizing territories with a high potential for socio-environmental risks.

Environmental and social problems are constantly observed in these areas. The big question is: Can social conditions be characterized as a determining factor in greater or lesser exposure to natural hazards? The initial answer to this question is that segregated urban areas usually have a simultaneity of these factors, making them more vulnerable to phenomena that cause human and material damage.

The territories of risk and vulnerability arise from the combination of factors such as: precarious housing conditions (which do not offer the minimum security to those who inhabit them due to the materials and engineering that were used to build them); lack of infrastructure that is the responsibility of the public authorities (drainage works, sewage, garbage collection) and the irregular occupation of spaces that have their own natural dynamics, which is the case of Permanent Protection Areas (APP).

The problem to be investigated is linked to the complex relationship between social vulnerability and areas prone to natural risks in cities, in this case in the capital of Rio de Janeiro. In other words, the thematic focus will be on social vulnerabilities and how they influence greater exposure to potentially damaging natural phenomena.

By mapping the risk areas of a given territory, prevention measures become easier to see due to the spatialization of these risks. By carrying out this study, it is possible to identify the type of risk, its spatialization and degree of vulnerability, in other words, it is possible to obtain the necessary information for planning risk areas, answering important questions such as: What is the risk? Who is vulnerable? What is the degree of

---

geological stability, biodiversity, the gene flow of fauna and flora, protecting the soil and ensuring the well-being of human populations. Thus, according to Chapter II, Section I, Article 4, the following are considered APP: forests and other forms of natural vegetation situated along rivers or any other watercourse from their highest level in a marginal strip; around lagoons, lakes or natural or artificial water reservoirs; at springs, even intermittent ones, and at the so-called "eyes of water", whatever their topographical situation; on the tops of hills, mountains and mountain ranges; on slopes or parts thereof with a gradient of more than 45°, equivalent to 100% on the line of greatest slope; on sandbanks, as dune fixers or mangrove stabilizers; on the edges of trays or plateaus, from the line of rupture of the relief, in a strip never less than 100 (one hundred) meters in horizontal projections; and at an altitude of more than 1.800 meters, whatever the vegetation.

vulnerability?

The city of Natal was basically built on the trays of the barrier formation surrounded by dune deposits; neodunes and paleodunes can easily be found in the city; coastal, fluvial and fluvio-marine plains such as along the Potengi river estuary, with the presence of mangrove forest (OLIVEIRA, 2003). The city does not have high altitudes, but the steepness of some slopes brings risks of landslides of unconsolidated material, mainly from sedimentary deposits such as dunes, which in turn are intensively occupied. Another common risk in the capital is flooding, which causes financial and health damage to the population exposed to contaminated water.

This study is based on the hypothesis that natural hazards and social vulnerabilities overlap in the city of Natal, which leads us to some questions such as: Is exposure to natural hazards greater for certain social groups? Who is most vulnerable in the city? Is there in fact a spatial juxtaposition between areas of natural risk and the most socially vulnerable areas in the city?

The main objective of the study was to map the areas of social vulnerability to exposure to natural hazards in the city of Natal (figure 01 delimits the study area), taking into account the interrelationships between social vulnerability and differentiated exposure to natural hazards. To this end, it will be necessary to establish, according to the methodology employed, the degree of social vulnerability and vulnerability to natural hazards to which individuals are subject, in order to establish the relationship between society and risks.

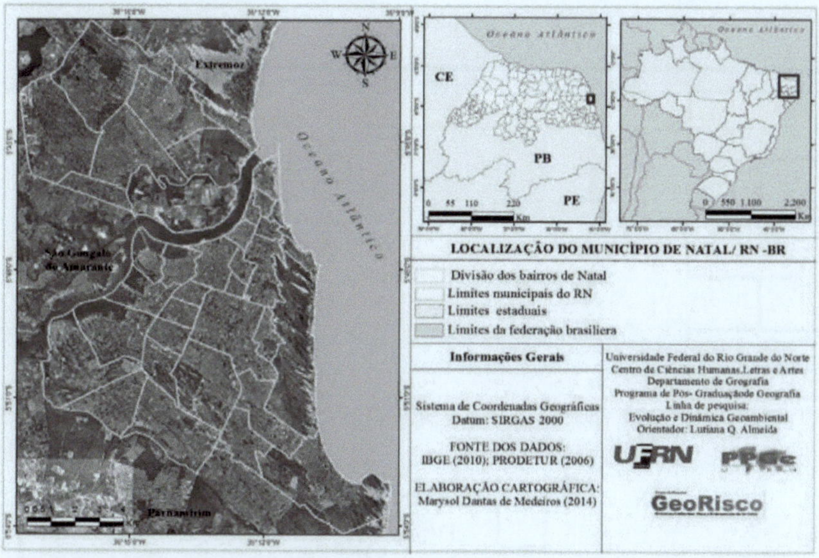

Figure 01: Location map of Natal/RN. **Source:** Prepared by Marysol Medeiros using data from PRODETUR (2006) and IBGE (2010).

In order to meet the objectives of this study, it was necessary to carry out a number of steps in accordance with the following methodological procedures set out in the flowchart (figure 02).

Firstly, bibliographical surveys were carried out on the subject to support the theoretical and methodological framework used for this study; in addition, consultations were carried out with newspaper libraries in order to obtain past evidence of episodes of natural disasters in the study area. After consolidating the theoretical framework, the Social Vulnerability Index (SVI) and the Environmental Vulnerability Index (EVI) were developed. The SVI was based on secondary data from the 2010 Demographic Census carried out by the Brazilian Institute of Geography and Statistics (IBGE); In order to compile the VAT, primary data was compiled, as well as secondary data from public bodies to help make the necessary thematic maps, so maps, topographical charts and tables with information about the physical environment of the capital of Rio de Janeiro were consulted. At the same time, fieldwork was carried out to check the thematic maps of both the IVS and the IVA, so that these maps could be superimposed and the IVSA could be produced.

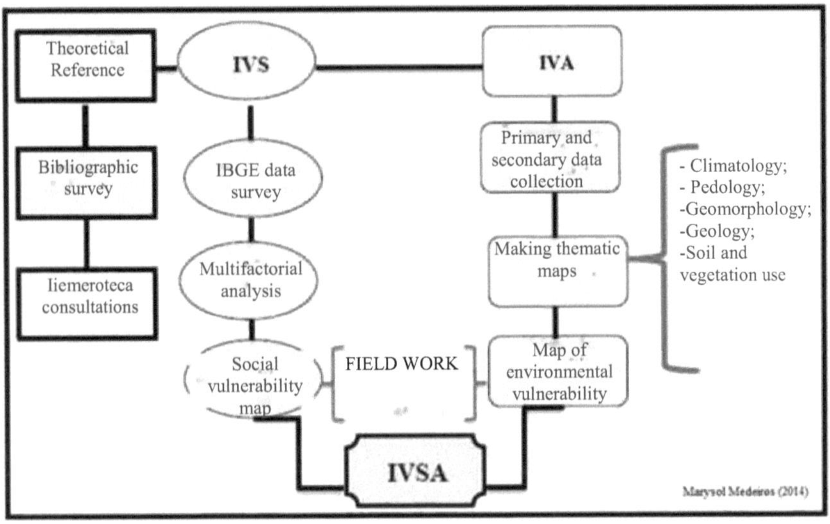

Figure 02: Flowchart of methodological procedures. **Source:** Prepared by Marysol Medeiros.

Vulnerability in Geography; it is necessary to define, among the various conceptions of risk, danger and Vulnerability, those that are closest to the debate of geographic science, as well as to trace the evolution of these concepts.

The second capital deals with the social vulnerabilities that contribute to the formation of risk territories in Natal. These vulnerabilities are responsible for the greater exposure of certain social groups to natural risks. This chapter presents Natal's Social Vulnerability Index (IVS), which was produced by compiling data by census tract from the Demographic Census carried out in 2010 by the Brazilian Institute of Geography and Statistics (IBGE), through a multifactorial analysis of this data.

The third chapter presents a diagnosis of the environmental factors that contribute to the occurrence of natural hazards in Natal, such as flooding and landslides. An exhaustive survey and analysis of secondary and primary data was carried out, producing thematic maps (geological, pedological, slope and altimetry, vegetation and land use maps) which were superimposed, culminating in the creation of the city's Environmental Vulnerability Index (IVA).

The fourth chapter deals with the results of the overlapping mapping of social and environmental vulnerabilities in Natal; bringing the Socio-environmental Vulnerability Index (IVSA), this chapter being the main objective of the research.

2. THE STUDY OF RISKS AND VULNERABILITIES FROM A GEOGRAPHICAL PERSPECTIVE

In order to meet the needs demanded by this study, we will initially look at the theoretical basis for the research, in this case, the Geosystemic approach, because as explained throughout this chapter, this approach is capable of encompassing both social and environmental factors with a view to understanding geographical space; However, the evolution of this approach in geographic science will not necessarily be discussed, but rather how it supports geographic studies in the light of the relationship between society and nature.

However, the main objective of this chapter is to present a theoretical discussion of the concepts of risk, danger and vulnerability, showing how they are incorporated into geographical science.

2.1 From the Geosystemic Approach to the conceptualization of Risk, Hazard and Vulnerability

In studies of urban areas that involve both social vulnerabilities and exposure to natural risks, it is of the utmost importance to understand the interactions between the dynamics of natural elements, as well as the social conditioning factors that expose an individual or group of individuals to a risk situation.

In order to find a suitable foundation for this study, which deals with the analysis of geographical space based on the relationships between natural dynamics and human actions, the theoretical framework to be used is based on Bertalanffy's General Systems Theory (GST), because, as Christofoletti (1979) states, a system is not merely the sum of its parts, and it is therefore essential to seek to understand a system from a holistic viewpoint; in this case, nature and its interrelationships with social dynamics. As Almeida (2010) points out, most of the socio-environmental problems in cities have to do with the lack of a systemic and holistic view of the elements that make up the urban space.

Until the mid-1960s, Physical Geography didn't have its own theoretical and methodological framework, having to resort to concepts from other natural sciences, mainly Ecology and its studies on ecosystems. For geographical studies, the General

Systems Theory was adapted, culminating in the Geosystemic approach, which is highlighted in this discussion. In the 1960s, the Russian Sotchava proposed the Geosystemic approach, in which geographers were able to classify homogeneous areas of natural landscapes by means of biogeographic criteria at the planetary, regional and topological levels, even though this concept was quite generalized. In 1968, the French geographer Georges Bertrand improved this approach by bringing it closer to the socio-economic scale, as noted by Nascimento and Sampaio (2004):

> In an attempt to synthesize the landscape, Bertrand establishes a taxonomic system for the geosystem, making it possible to classify it according to scale, characterizing it as a unit, a taxonomic level in the categorization of the landscape, namely: the zone, the domain and the region, as higher units, and the geosystem, the geofacies and the geotope, as lower units; the geosystem being provided by the dynamic between ecological potential, biological exploitation and anthropic action. This allows it to be placed in the dual perspective of time and space, which are fundamental to the geographer (NASCIMENTO; SAMPAIO, 2004, p.169).

As Nascimento and Sampaio also point out, "geosystems are a good basis for geographical studies because they are on a scale compatible with the human scale" (NASCIMENTO; SAMPAIO, 2004, p.171). As the following diagram by Georges Bertrand shows, the Geosystemic Approach provides support for the study of geographical space based on the interactions between social action, ecological potential and biological exploitation.

Tricart also explains that Bertrand's conception of geosystems *"distinguishes three successive levels: the physical environment, eco-systems, human intervention. This makes it possible to introduce, without difficulty, a temporal dimension, a dynamic and evolutionary perspective"* (Tricart, 1979, p. 36).

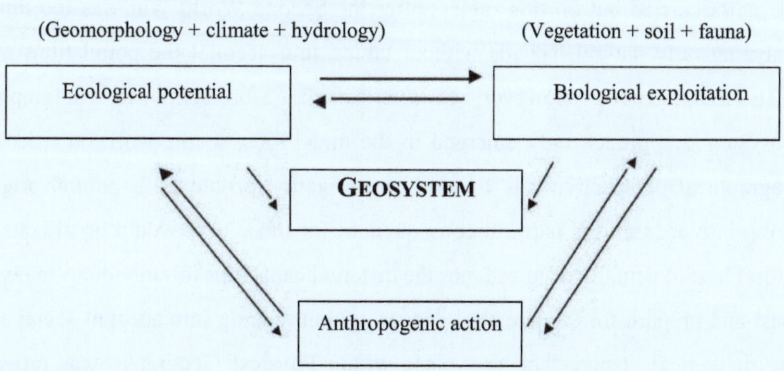

Figure 03. Outline of a theoretical definition of Geosystem. **Source:** BERTRAND (1972).

From the geosystemic point of view, geographers can direct their environmental studies in a way that integrates, above all, the social dynamics that interfere, directly or indirectly, with nature, so that nature is no longer approached in geography under the aegis of ecology, in which the ecosystem is governed by exceptionally natural laws. Rodrigues (2001) points out that the geosystemic approach has brought a dynamic character to the study of the landscape, since it is possible to classify geosystems by forming a historical series according to the state in which the landscape was in different periods, thus enabling prognoses to be made with a view to territorial planning, which is still the focus of this research.

Studies of natural hazards, or "*natural harzard*", among geographers date back to the 1920s, with 1927 being considered a milestone for this line of research according to Gilbert F. White - considered one of the most important authors on the subject - because it was during this period that the United States government asked the U.S. government to solve flooding problems in its territory. White - considered one of the most important authors on the subject - because it was during this period that the United States government, in order to solve the problems of flooding in its territory, asked the *U.S. Corps* of Engineers to research solutions (MARANDOLA; HOGAN, 2005). It should be noted that, for a long time, this subject was dealt with in the field of Geography, above all by physical geographers, who in turn focused on researching the physical and natural aspects, their processes, their chronologies and their measurement (ALMEIDA, 2011).

The work carried out on this subject after the Second World War was too limited to analyzing only natural hazards without taking into account the populations affected (AZEVEDO, 2010). However, as seen in the aforementioned paragraphs, the geosystemic approach only emerged in the mid-1960s, so the work on risks from a geographical perspective was focused on mitigating problems of natural origin, not taking into account the harmful consequences for those who would be affected if the natural hazard actually occurred, nor the different capacities of individuals to deal with, resist and prepare for the potential danger, without taking into account social aspects. The theoretical change that took place within Physical Geography was reflected in studies on "*natural hazards*" when, as Almeida puts it, "there was a growing awakening to the interest in the relationship between human activities and the environment" (ALMEIDA, 2011, p. 86).

> It was also in the mid-1970s that a trend towards a more conjunctive approach to *natural hazards* began to emerge. At the time, there was an approach to *hazards* in Geography, based on its inheritance from Human Ecology, with an emphasis on the relationship between natural hazards and social responses to them and on reducing losses (MILETI, 1999). Coming from Sociology, an approach emerged whose perspective was the analysis of collective behavior and emphasized the response to disasters and their prevention. This theoretical orientation became known as "*disaster research*" (ALMEIDA, 2011, p.87).

The changes that took place in society, especially after the Second World War, with the intensification of capitalism and industrialization culminating in widespread urbanization, deepened social and environmental problems such as growing social inequalities and environmental degradation. In view of this situation, a new theoretical-methodological approach emerged within the Geography of Risks in the 1980s, which sought to focus especially on the populations affected (ALMEIDA, 2010). Thus, Castro (2000) shows that studies on risks in the 1980s began to emphasize the relationship between natural hazards and economic underdevelopment, since the latter increases the damaging effects of disasters, also causing greater attention to anthropogenic hazards, and the recognition of the multidisciplinarity of the subject.

As Bertrand (2007) puts it, geographers are caught between discussions about naturalistic knowledge and certain forms of social analysis. It is because of this interdisciplinary capacity and the dichotomous analysis between society and nature that geographers have incorporated (or been incorporated into) studies on natural risks. It should be stressed that, according to Azevedo (2010), regardless of the "branch" of Geography, geographers are of fundamental importance in carrying out research related to natural hazards, given their multidisciplinary training, as well as their ability to spatialize such risk phenomena through cartography. In a similar vein, Marandola and Hogan (2006) highlight the role of geographers in studies on natural hazards:

> This tradition is important because, unlike social scientists who focus on the vulnerability of people or groups, geographers tend to be concerned with the vulnerability of place (CUTTER, 1996). This approach starts from the dynamics that shape a given spatiality, seeking to circumscribe its scale (a region, a city, an ecosystem, a neighborhood), identifying in the interactions between society and nature the risks and dangers that affect the place. It is not a question of understanding this spatiality as a physical

substrate independent of society. Rather, the approach seeks a reference unit to understand the context of the social production of danger in connection with the geographical context in the delimitation of the spatial scale (MARANDOLA; HOGAN, 2006, p. 36).

Geographers are well aware of the importance of studying risks, especially natural risks. However, what is risk? Which types of risk fall within the scope of geographic studies? In order to answer these questions, the definitions of risk, danger and vulnerability will be elucidated, since these three terms are translated by common sense as synonyms, but they are secondary concepts for studies on risks and vulnerabilities in the field of Geography, given that they are strongly linked.

2.1.1 Defining risk and danger

Various sciences turn their studies to analyzing risks, whether they are economic, natural, related to health and safety, among others; and this word is always linked to something negative, since, for today's model of society, taking risks is something extremely counterproductive, due to the varied possibility of predicting certain phenomena or changes of the most diverse kinds, avoiding possible damage. However, this negative connotation of risk has been different in other times, being a neutral term that can refer to good consequences or not (MARANDOLA, 2005).

In the past, the term risk was related to the possibility of going and not returning from long voyages across unknown oceans. As Rebelo (2010) reports:

The word risk is first mentioned in connection with life at sea. [...]. In fact, when sailors set off, they didn't know if they would reach their destination, let alone if they would return. That's why they began to use a word to signify the various problems they faced. It was the risk of going to sea. Soon, risk began to be linked to cargo insurance. From then on, risk and insurance went hand in hand (REBELO, 2010, p. 32).

The origin of this term is not well defined, but in many works, its etymology linked to the Great Navigations of the 15th century is quite recurrent. There are also other possible origins for it, as elucidated by Castro (2000), meaning struggle, division, contradiction; linked to geomorphology as a synonym for steep cliff. The same author recalls that the origin of the term risk comes from *risico* or *rischio* (danger).

The origin of the term riesgo is uncertain; according to Diez and other linguists, it is related to the old Castilian *resegue* (resecar, cortar), whose meaning, widely used in the Middle Ages, is synonymous with struggle, contradiction and division. It is therefore thought that the whole *riesgo-risco* group probably comes from the Latin *resecare*, to cut, which has a double meaning: on the one hand, division, discord and on the other, a broken and fragile place. Etymologically riesgo comes from *risico* or *rischio* (danger). It is believed that it may have a common origin with the Castilian word '*risk*': steep cliff, cliff, promontory, formerly, risk, which was also applied to the danger faced by those who passed through steep cliffs or promontories (*rhizicare*) (CASTRO, 2000, p. 2).

It is necessary to understand by definition what risk means. In post-modern times, the term has come to mean something bad. But what is its definition? From a few definitions of this concept, it will be possible to glimpse its real meaning in science, especially in the light of Geography.

Thywissen (2006) has compiled a glossary of various terms related to natural disasters, including definitions of risk, hazard and vulnerability. For the term risk, this glossary presents fifteen different definitions for different areas of knowledge, as shown in Table 01.

Table 01: Glossary with definitions of the term risk.

Definition	**Source/ Discipline**
(In this definition risk and hazard are used synonymously) "Risk is characterized by knowing or not knowing the probability distribution of events. These events are characterized by their magnitude (including size and spread), their frequency and duration, and their history."	Alwang; Siegel; Jorgensen (2001) / Social Sciences
"Risk: the expected number of lives lost, people injured, damage to property and interruption of economic activity due to a given natural phenomenon, and consequently the specific product of risk and elements at risk. " "Thus, risk is the potential for loss for the exposed subject or system, resulting from the 'convolution' of danger and vulnerability. In this sense, risk can be expressed in mathematical form as the probability of overcoming certain economic, social or environmental consequences in a certain place and over a certain period of time."	Cardona (2003) / Science (multidisciplinary)
"Risk is a function of the specific probability of the natural hazard event and the vulnerability of cultural entities. "	Chapmann (1994) / Natural Sciences
"Risk can be defined as the probability of a system not being in a satisfactory state. "	Correia; Santos; Rodrigues (1987) / Engineering
"'Risk' is the probability of a loss, and this depends on three	Crichton (1999) / Natural Sciences/

elements: danger, vulnerability and exposure. If any of these three elements of risk increases or decreases, then the risk increases or decreases respectively. "	Insurance sector
"Risk is 'the probability of an event multiplied by its consequences if the event occurs'."	Einstein (1988) / Natural Sciences
"A combination of the probability or frequency of occurrence of a defined hazard and the magnitude of the consequences of the occurrence. More specifically, a risk is defined as the probability of a harmful consequence, or expected losses (of life, people, injuries, property, livelihood, disrupted economic activity or deteriorated environment) resulting for interactions between natural or human induced hazards. "	European Spatial Planning Observ.Netw. (2003) / Science (multidisciplinary)
"The risk associated with a flood disaster for a region is a product of both the exposure to the hazard (natural event) and the Vulnerability of objects (society) to the hazard. This suggests that three main factors contribute to a flood risk region: hazard, exposure and vulnerability. "	Hori et al. (2002) /Geosciences
"Objective (mathematical) or subjective (inductive) probability that the hazard will become an event. Factors (risk factors) can be identified that interfere with this probability. Such risk factors are made up of personal behaviors, lifestyles, cultures, environmental factors, and inherited characteristics that are known to be associated with health issues. Risk is the probability of loss of the elements at risk as a result of the occurrence, physical and social consequences of a natural or technological hazard, and the mitigation and prevention measures in place and in the community. Risk is the expected number of lives lost, people injured, property damaged and economic activity interrupted due to a particular natural phenomenon, and consequently the specific product of risk and elements at risk. " UNDRO	Journal of Prehospital and Disaster Medicine (2004)/ Science (multidisciplinary)
"Risk indicates the degree of potential losses in an urban site due to its exposure to hazards and can be considered as a product of the probability of hazards occurring and the degree of vulnerability. "	Rashed; Weeks (2003) / Geosciences
"The probability of exposure to an event, with which it can occur with varying severity and different geographical scales, suddenly and unexpectedly or gradually and predictably, and the degree of exposure. "	UNEP (2002) / United Nations

"Used in an abstract sense to indicate a real-world condition in which there is a possibility of loss; also used by insurance professionals to indicate property insurance or insured peril. "	Swiss Re (2005) / Insurance sector
"Risk is the expected number of lives lost, people injured, property damaged and economic activity interrupted due to a particular natural phenomenon, and consequently the specific product of risk and elements at risk. Specifying risk: the degree of loss expected due to a particular natural phenomenon and a function between both natural hazard and vulnerability. "	Tiedemann (1992) / Insurance sector
"The probability of harmful consequences, or expected loss of life, injury, property, livelihood, interruption of economic activity (or damage to the environment) resulting from joint interactions between natural and human inducing conditions of danger and vulnerability. Risk is conventionally expressed by the equation: Risk = danger x vulnerability. "	UNDP - BCPR (2004) / United Nations
"System risk can be defined simply as the possibility of an adverse and undesirable event. Risk can be due solely to physical phenomena such as a health hazard or to the interaction between artificial systems and natural events, such as a flood due to the breach of a dyke. Engineering risk for water resource systems in general has also been described in terms of a figure of merit that is a function of performance indices, say, for example, reliability, incident period, and repair possibilities..."	Shrestha (2002) / Engineering

Source: Thywissen (2006) adapted by the author

However, it is possible to see from the following definitions that "risk" is related to the probability of a certain phenomenon occurring or not, affecting in a harmful way, which can cause damage of the most diverse kinds (damage to health, life, property, the economy, the environment, among others).

Risk is the likelihood of a potentially harmful event occurring, but it is a social construct which involves the perception of exposed individuals and their vulnerabilities to danger. Following Veyret's (2007) definition, it is "the perception of a possible danger, more or less predictable by a social group or by an individual who has been exposed to it" (VEYRET, 2007, p. 24).

In simplified terms, risk can be considered to be linked to an event that may or may not happen. However, the existence of a risk is only constituted when there is a valuation of some asset, material or immaterial, because there is no risk without the notion that something could be lost. Therefore, one cannot think of risk without considering someone who is at risk, that is, society (CASTRO; PEIXOTO; RIO, 2005, p. 27).

Risk is seen as a human perception, so an individual or a group of individuals can be considered vulnerable to an event that could cause damage, whether to human integrity or to material goods. In this way, there is no risk if man is not present (directly or indirectly) in the face of a given danger, because as Rebelo (2010) explains, "if man is not present, it is often considered that it is not legitimate to speak of risk. For there to be risk, it is said, there must be vulnerability" (REBELO, 2010, p. 32).

It should be emphasized that risk is not only linked to the probability of a potentially harmful event occurring, paraphrasing Chardon (1998) who quotes David Le Bretón (1995), risk is also a socially constructed notion, eminently variable from one place and time to another, since each social and cultural condition assumes its own fragilities, feeding a cartography of what is feared.

The same author also states that three levels of risk can be distinguished, which are established on the basis of a set of variables that make a population more or less exposed to a danger, or rather, they are levels that indicate the tolerance capacity of a potentially exposed population. Thus, a tolerable risk would be one in which the degree of exposure to the danger or the vulnerability of the population is lower; the risk can also become tolerable after technical interventions that reduce the frequency and magnitude of the danger, as well as the preparation of the population in advance so that, when exposed, they know how to deal with the event; and finally, the risk can be intolerable when society is unable to resist or absorb the manifestations and consequences generated by the natural phenomenon.

Today's society is exposed to the most diverse types of risk, although various sciences use the term "risk" as a secondary concept in their studies, but for Veyret, despite the various types of risk, Geography should only deal with those that can be spatialized, i.e. those in which perception and management are followed by a spatial dimension; these, in turn, are classified according to the processes that conceive them, as shown in table 02 organized by Almeida (2010).

Chart 02: Types of spatialized risks.

Types of Risks		Definitions, characteristics, examples
Environmental Risks	**Natural Risks**	Risks sensed, perceived and borne by a social group or subject to the action of a natural physical process; they can be of lithospheric origin (earthquakes, landslides, volcanic eruptions), and hydroclimatic (cyclones, storms, heavy rains, floods, blizzards, hailstorms, droughts); they have physical causes and largely escape human intervention and are difficult to predict.
	Natural risks aggravated by man	The result of a natural hazard whose impact is magnified by human activities and land use; erosion, desertification, fires, pollution, floods, etc.
Technological Risks	They are distinguished into chronic pollution (dangerous phenomena that occur on a recurring basis, sometimes slowly and diffusely) and accidental pollution (explosions, leaks of toxic products, fires).	
Economic, geopolitical and social risks	Risks linked to the division of and access to certain resources (renewable or not), which can translate into latent or open conflicts (as in the case of oil and water reserves); they can also originate from economic relations in agriculture (food insecurity), causes of globalization (economic crises), insecurity, violence due to urban socio-spatial segregation, health risks (epidemics, hunger, pollution, drug use, etc.).	
Other Types of Risks	**E.g.: Risks More**	The understanding of risk also depends on the scale of analysis; a major risk is thus considered when the cost of recovery and the number of human losses are relatively high for public authorities and insurers; major risks correspond to events of low frequency and great magnitude and consequences (e.g.: Chernobyl, Seveso, Bhoppa, Katrina, etc.).There are also examples of the "thematization" of risks, such as the specific case of urban risks, due to the complexity and multidimensionality of actors and variables in cities.
	E.g. urban risks	

Source: Organized by Almeida (2010), based on Veyret (2007).

According to the table, the term risk is associated with an adjective that qualifies it, but all of these risks have in common their spatialization in a territory, which is the main reason why they are addressed by Geography. Not just taking into account a Ratzelian view of territory, today there is a tendency to analyse risks in geographical science based on their territorialization, as can be seen from publications such as those by the Frenchman Robertier (2012), in which he argues that physical factors play a fundamental role in the formation of environmental risks, however, the structure and implications of this risk will depend heavily on social organization, power relations and social inequalities. For the same author, the concept of risk is defined as a "social construction involving a wide range of variables that depend on the time, place and society" (ROBERTIER, p. 392, 2012).

The term danger is often confused with risk, as if they were synonyms, which is a mistake. Danger is the phenomenon that causes damage or harm. For Veyret (2007), "this term is sometimes also used to define the objective consequences of a[3] hazard on an individual, a group of individuals, the organization of the territory or the environment. A potential and objective fact" (VEYRET, 2007, p. 24).

Hazards can have even greater consequences for an individual, or group of individuals, depending on their magnitude. An example of this was what happened in Japan in 2011, when a *tsunami* formed in the Pacific Ocean after an earthquake decimated countless cities along the Japanese coastline, claiming the lives of thousands of people and causing incalculable economic damage. However, it is worth noting that this is a country known for its ability to constantly deal with natural hazards due to its geography, which is conducive to such phenomena; however, due to the exceptional magnitude of the event, the population was unable to withstand it.

Called *Hazard* in English-speaking countries, danger is defined by Smith as a process of natural or *human-induced* origin with the potential for future loss: "*hazard is best viewed as a naturally occurring or human-induced process, or event, with the potential to create loss, that is, a general source of future danger.*" (SMITH, 2001, p. 6)

Cardona (2003), in his definition of *hazard,* shows a similarity between this concept and that of risk. For him, *natural hazard* would be the probability of a potentially damaging phenomenon occurring over a specific period in a given area.

3 According to (Veyret, 2007, p.24) àlea would be the "possible event; it can be a natural, technological, social or economic process and its probability of realization." If several events are possible, we speak of a set of àleas. The equivalent in English is hazard and in Portuguese the term is "perigo".

As you can see, risk and danger have similar definitions; not only that, but the term risk comes from the term danger, hence the controversy surrounding them. Remembering that risk is linked to the probability of danger occurring, however, for there to be risk and danger there must be something or someone exposed, otherwise there will be no probability of damage or loss occurring, a factor so often repeated in current definitions of risk. As risk is linked to probability, what determines this probability is the degree to which something or someone is exposed to danger; this measurement is therefore called vulnerability.

Next, we will discuss the concept of vulnerability, a concept that was incorporated into *natural hazard* studies at a late stage, but which has completely changed the way we deal with, mitigate and manage crises arising from natural events.

2.1.2 Defining Vulnerability

Until the 1970s, studies on *natural hazards* were focused only on issues related to mitigating the possible consequences of the occurrence of a natural hazard. However, the changes that have taken place within society have forced a theoretical and methodological change in these studies, because the view centered only on immediate solutions to physical-natural factors did not encompass in a broader way the elements that trigger crises from natural hazards, and it was necessary to incorporate the perspective of the populations affected (ALMEIDA, 2010). Thus, especially in the 1980s, the concept of Vulnerability was added to these studies, emerging as a point of fusion between the social, cultural and environmental dimensions.

In the 1990s, the concept of Vulnerability became widespread in studies on natural hazards, as it gives a more holistic character to the elements at risk and can analyze, above all, the harmful potential of the hazard, i.e. who, or what, is more likely to suffer from the harmful impacts of a hazard due to their physical, social and cultural conditions. This makes it possible to infer more in-depth measures to prevent natural disasters, which are not necessarily more economically costly, since many problems could be solved simply by educating individuals globally to learn how to prevent, deal with and overcome the consequences of a hazard.

Like the concept of risk, the concept of vulnerability is also used by various fields of knowledge, ranging from the human sciences to the natural sciences - in geography, anthropology and sociology, engineering, ecology, among others. However, the difficulty of defining "vulnerability" is not only a matter of epistemological diversification among the sciences that adopt this concept, but also the difficulty in grasping the multidimensionality of the reality analyzed, which interferes with the operationalization of this concept (ALMEIDA, 2010). Thus, there are various definitions of vulnerability, given its complexity, since it engenders various nuances that make an individual or a group more exposed, i.e. more vulnerable to a given danger, which were not previously taken into account in studies of natural hazards.

Cutter (1996) draws attention to the multiplicity of definitions of vulnerability between the 1980s and 1990s. As can be seen in the following table (Table 03), the definitions of this concept have evolved in such a way that it used to be defined as the threat of exposure to hazardous materials occurring; which subsequently became more complex definitions involving the living conditions of exposed individuals in order to cope with and recover from the natural disaster.

Table 03. Selection of definitions of Vulnerability according to Cutter (1996)

AUTHORS	DEFINITIONS
Gabor; Griffith (1980)	Vulnerability is the threat (for hazardous materials) to which people are exposed (including chemical agents and the ecological situation of communities and their level of emergency preparedness). Vulnerability is the context of the risk.
Timmerman (1981)	Vulnerability is the degree to which a system acts adversely to the occurrence of a hazardous event. The degree and timing of adverse reactions are conditioned by a system's resilience (a measure of the system's ability to absorb and recover from the event).
UNDRO (1982)	Vulnerability is the degree of loss of a certain element or set of elements at risk as a result of the occurrence of a natural phenomenon of a certain magnitude.
Susman et al (1984)	Vulnerability is the degree to which different classes of society are differentially at risk.
Kates (1985)	Vulnerability is the 'capacity to suffer damage and react adversely'
Pijawka; Radwan (1985)	Vulnerability is the threat or interaction between risk and prevention. It is the degree to which hazardous materials threaten a particular population (risk) and the community's ability to reduce the risk or adverse consequences of the hazardous materials released.
Bogard (1989)	Vulnerability is operationally defined as the inability to take effective action against losses. When applied to individuals, vulnerability is a consequence of the impossibility or improbability of effective mitigation and is a function of our ability to detect danger.

Mitchell (1989)	Vulnerability is the potential for loss.
Liverman (1990a)	She distinguishes between vulnerability as a physical condition and vulnerability as defined by the political, social and economic conditions of society. She argues for vulnerability in geographical space (where the most vulnerable people and places are located) and vulnerability in social space (which place is vulnerable).
Downing (1991b)	Vulnerability has three connotations: it refers to a consequence (e.g. hunger) rather than a cause (e.g. they are vulnerable to hunger), and it is a relative term that differentiates between socio-economic groups or regions, rather than a measure of deprivation.
Dow (1992)	Vulnerability is the differential capacity of groups and individuals to deal with risks, based on their positions within the physical and social worlds.
Smith (1992)	Specific hazard risk varies over time and according to changes in either (or both) physical exposure or human vulnerability (the range of social and economic tolerance available in the same place).
Alexander (1993)	Human vulnerability is a function of the costs and benefits of living in areas at risk of natural disasters.
Cutter (1993)	Vulnerability is the probability of an individual or group being exposed to and adversely affected by a hazard. This is the interaction of local hazards (risk and mitigation) with the social profile of communities.
Watts; Bohle (1993)	Vulnerability is defined in terms of exposure, capacity and potential. Therefore, the prescriptive and normative response to vulnerability is to reduce exposure, increase coping capacity, strengthen recovery potential and reinforce damage control (minimizing destructive consequences) through public and private means.
Blaikie et al. (1994)	By vulnerabilities we mean the characteristics of a person or group in terms of their ability to anticipate, cope with, resist and recover from the impact of a natural hazard. This involves the combination of factors that determines the degree to which one's life and livelihood are put at risk by a natural hazard.
Bohle et al (1994)	Vulnerability is best defined as a global measure of human well-being that integrates environmental, social, economic and political exposures to a range of potential harmful disturbances. Vulnerability is a space of several social and multidimensional layers defined by the determined capacities, political, economic and institutional resources of people in specific places and at specific times.
Dow; Downing (1995)	Vulnerability is the differential susceptibility of circumstances that contribute to vulnerability. Biophysical, demographic, economic, social and technological factors, such as population age, economic dependence, racism and age of infrastructure are some of the factors that have been analyzed in association with natural hazards.

Source: Cutter (p. 531-532, 1996); adapted and translated by Marysol Medeiros.

Despite this diversity of definitions, Thywissen (2006), in his glossary, provides a further 28 definitions of the concept of vulnerability, and those that have a certain affinity with this work are listed below.

Table 04. Selection of definitions of Vulnerability according to Thywissen (2006)

Definition	**Source/ Area of knowledge**
Vulnerability (in contrast to poverty, which is a measure of the current situation) must involve a predictive quality: it is supposed to be a way of conceptualizing what can happen to an identified population in particular conditions of risk and danger. It is the complex set of characteristics that include an individual - Initial well-being (health, morale, etc.) - Self-protection (standard assets, income, qualifications, etc.); - Social protection (preparing society for danger, building codes, shelters, etc.) - Social and political networks and institutions (social capital, institutional environment).	Cannon, Twigg and Rowell (2003) Social Sciences
Vulnerability.... It's not the same as poverty. It means not having or wanting, but lacking protection, insecurity and exposure to risks, shocks and stresses... Vulnerability refers to exposure to contingencies and stress, and difficulty in dealing with them.	Chambers (1989) Social Sciences
Vulnerability expresses the seriousness of the failure in terms of its consequences. The concern is not how long the failure lasts, but how costly it is.	Correia, Santos and Rodrigues (1987) Engineering
It is a broad measure of susceptibility to loss or damage. The greater the vulnerability, the greater the exposure to loss or damage.	Department of Human Services (2000) Social Sciences
"The degree of loss of a given element at risk (or a set of elements) resulting from a given hazard at an assigned level of severity." In contrast to the concept of risk, here the probability of a hazard occurring is not taken into account (UNDP-BCPR/UNDHA, 1994:38-39, see also UNDHA, 1992). Vulnerability is a process and is not static.	Feldbrugge and von Braun (2002) United Nations
Vulnerability (V) = Hazard - Ability to cope with Risk = H (Probability of hazard or process, shock value, predictability, prevalence of intensity/strength), AND coping capacity = C (Perception of risk and potential of an activity possibilities of trade, private trade, open trade).	Feldbrugge and von Braun (2002) p.11 Science (Multidisciplinary)
Determinants of vulnerability to disasters: - Demographic factors: population growth, urbanization, settlements near coastal areas, etc;	Feldbrugge and von Braun (2002, p.14) Science

- The state of economic development: poverty, modernization processes; - Environmental changes: climate change, degradation and depletion of resources (changes in river courses, deforestation, etc.) - Political factors, an increase in tangible assets, which leads to an increase in damage, the effects of disaster protection structures and research, and the interactions of the causes of disasters.	(Multidisciplinary)
The probability that some defined social group in society will suffer disproportionate death, injury, loss or interruption of livelihoods in an extreme event, or face greater than normal difficulties in recovering from a disaster.	Handmer and wisner (1998) Science (Multidisciplinary)
The characteristics of a person or group in terms of their ability to anticipate, cope with, resist and recover from the impact of a natural or man-made hazard.	IFRC (1999) Disaster management
Vulnerability is provisionally defined as the degree to which a system is sensitive to and unable to cope with the negative effects of global change stimuli. Vulnerability is therefore a function of a system's exposure to global change stimuli and its adaptive capacity, i.e. its ability to cope with these stimuli.	Klein (2003) Science (Multidisciplinary)
Vulnerability defines the inherent fragility of certain aspects of the urban environment that are susceptible to damage due to social, biophysical or design characteristics.	Rashed and Weeks (2003) Science (Multidisciplinary)

Source: Thywissen (2006) adapted by Marysol Medeiros.

However, for this study, vulnerability would be the measurement of each individual's ability to prepare, cope and resist when exposed to danger. "Vulnerability measures the harmful impacts of the event on the affected targets" (VEYRET, 2007, p. 24). We also agree with Blaikie et al's (1994) definition, which takes into account the characteristics of the individual or group to anticipate, cope with, resist and recover from a risk situation.

By vulnerability we mean the characteristics of a person or group and their situation that influence their capacity to anticipate, cope with, resist and recover from the impact of a natural hazard (an extreme natural event or process). [...] Some groups are more prone to damage, loss and suffering in the context of different hazards (BLAIKIE et al, 1994, p. 11).

Vulnerability will exist from a danger in a given geographical and social context, as there are places more prone to the occurrence of harmful phenomena than others, as well as individuals who, due to various factors - social, political, cultural, etc. - are less resilient. The time factor must also be taken into account, given the dynamic nature of vulnerability, since the factors that expose something or someone are extremely changeable, and can be seasonal and appear on a wide range of time scales (MARANDOLA; HOGAN, 2006).

According to Chardon (1998) there are many factors that make an individual or group vulnerable, influencing the degree of exposure individually or not, where these factors can act alone or together as a system.

The vulnerability factors belong to different fields (natural, material, social, functional, decision-making, etc.) and influence not only considering them individually, but also in interaction with each other, thus forming a system. Complexity increases to the extent that it is possible to distinguish between two forms of risk: potential (or virtual) risk of which only specialists are aware; and declared risk, which reveals the dangerous situation for everyone (CHARDON, 1998, p. 3).

Faced with this controversy of definitions and factors that lead to vulnerability, Cutter (1996) states that studies on vulnerabilities have basically presented three biases, one linked to conceptual issues, with discussions about the concept and its definition; another that focuses on the different methodologies for measuring the various types of vulnerability; and finally, the factors that lead individuals to become more vulnerable to certain dangers. She also brings up a new concept on the subject (*hazard of place)* which encompasses vulnerability as potential exposure to danger, social response and the combination of these two in a given geography, i.e. vulnerability is the interaction between the potential risk of a given place and the social characteristics together with the degree of exposure of the individuals in that place; "*In this perspective, vulnerability is conceived as both a biophysical risk as well as a social response, but within a specific areal or geographic domain*" (CUTTER, 1996, p. 533). 533).

Almeida (2011) warns that in order to form a vulnerability system, attention must be paid not only to the type of risk - which is infinite - but also to the vulnerable elements and, above all, the factors that make these elements vulnerable.

The vulnerability system is made up of vulnerable elements and factors that make these elements vulnerable in the city. We can talk about factors linked to the problem of demographic and urban growth and its acceleration, including the forms of use and occupation of urban land; socio-economic factors (rural exodus, real estate speculation); psychosociological factors (memory of risk, perception and culture of risk); factors linked to the culture and history of exposed societies (self-building, waste disposal); technical factors (prevention); functional factors (crisis management); institutional factors (risk management); among other factors, such as the insurance system (ALMEIDA, 2011; p. 76). 76).

Just as there are various types of risk and various factors that make elements more vulnerable, there is also a range of types of vulnerability. Therefore, the following table (table 05) compiles some types of vulnerability.

Table 05: Types of vulnerability applied to studies of natural phenomena

Types of Vulnerabilities	Features
Physical (or structural, or bodily) vulnerability	They focus on analyzing buildings, infrastructure networks and the potential for human losses.
Human or social vulnerability	It assesses the feedback of experience on response capacities, adaptations, behaviors and their socio-economic and territorial consequences. It also includes the perception of threats or the memory of risk, knowledge of means of protection, and types of potential behavior.
Institutional vulnerability	It deals with the capacity of institutions to respond to the crisis; it works as an indirect factor of social vulnerability.
Environmental and property vulnerability	It analyzes the damage to environmental components - vegetation, soils, water resources, fauna and cultural aspects - caused by natural phenomena.
Functional and economic vulnerability	It assesses dysfunctions in economic activities, disruptions in communication and transportation networks, among others.

Source: ALMEIDA (2010), p. 110; based on Leone and Vinet (2006)

When analyzing social vulnerabilities from the perspective of geographic science, it is necessary to bear in mind that they occur in a particular place. Almeida (2010) states that social vulnerabilities are "constituted by social and spatial inequalities and, due to this characteristic, it is essential to evaluate and compare vulnerabilities between different spaces" (ALMEIDA, p. 157, 2010).

The same author, paraphrasing Cutter (2003), states that from the point of view of environmental sciences, vulnerability studies "can contribute to understanding the circumstances that put people at risk and the conditioning factors that reduce the ability of people and places to respond to environmental threats, i.e. reduce their resilience" (ALMEIDA, p. 3, 2010).

This work is related to the analysis of the overlapping of areas with high social and environmental vulnerability, which for Alves (2009) would be areas of Socio-environmental Vulnerability, since this term is defined as "the coexistence or spatial overlap between very poor population groups with high deprivation (social vulnerability) and areas of risk or environmental degradation (environmental vulnerability); It is precisely the combination of these two dimensions - social and environmental" (ALVES, 2009, p. 82). 82).

It was necessary to define the concepts of risk, danger and vulnerability, as these three are intrinsically linked, since there is only risk if there is a danger (which is nothing more than the phenomenon itself) and the degree of vulnerability of the individuals exposed. Thus, this statement leads to the conclusion that risk is a function between the danger that is potentially harmful and causes damage and the factors that make individuals more or less exposed to this danger (vulnerability). This sentence can be summarized in the following equation: R= f (P; V), where Risk (R) is a function between Danger (P) and Vulnerability (V).

The inclusion of vulnerability in risk studies has significantly expanded the way in which crises arising from extreme events can be predicted and managed, even though this concept does not have a consensual definition among the sciences that use it, since it makes it possible to investigate the preponderant causes for the occurrence of certain risk situations, which initially may not be related to natural danger, but which nevertheless increase the damage caused by the natural phenomenon.

2.2 Ecodynamics and Environmental Vulnerability

Crepani (2001), the author of the methodology on which this work is based, considers areas of environmental vulnerability to be those in which there is the greatest loss of soil, i.e. his methodology is based on Tricart's Ecodynamics (1977), which takes into account morphogenetic and pedogenetic processes, since soil loss is generally the result of mass movement, taking with it everything that is above this substrate, be it buildings, plantations, roads, among others; consequently, causing various types of damage.

In this way, Tricart (1977) takes into account the dynamism of the elements that make up the environment when classifying areas according to morphogenesis, because for him "Human action is exercised in a changing nature, which evolves according to its own laws, from which we perceive, more and more, its complexity" (TRICART, 1977, p. 35).

Thus, Tricart (1977) provides taxonomic classifications of the environment on three levels according to morphodynamics, i.e. according to the intensity of the current processes in the ecosystem. The taxonomic classification suggested by Tricart is as follows:

- Stable environments - morphodynamic processes evolve slowly and are only perceptible through precise measurements; these conditions are characteristic of regions with low internal geodynamic intensity and low intensity of the mechanical processes of external geodynamics, so pedogenesis takes precedence over morphogenesis.

- *Intergrades* - "These ensure the gradual passage between stable and unstable environments. In this environment, pedogenesis and morphogenesis take place concurrently in the same space" (TRICART, 1977 p. 47).

- Strongly unstable environments - there is a predominance of morphogenesis in the dynamics of these environments, and morphogenesis can occur both from endogenous sources, such as tectonic movements and volcanism; and from exogenous sources, such as the actions of weathering, winds, surface water run-off, among others; and from anthropogenic sources, mainly through environmental degradation.

In this way, these means (or taxonomic levels of the degree of morphogenesis) serve as the basis for establishing the degree of environmental vulnerability in the methodology proposed by Crepani (2001), as shown in Chapter 3 of this work.

2. 3 Risks and social vulnerability in cities

Cities are the *locus* for the installation of territories of risk and vulnerability, given that the global population is increasingly moving to urban areas. However, many cities can no longer cope with the influx of people attracted by various factors.

Failing to find suitable areas for housing, many people are forced to live in places that have intense natural dynamics and which are therefore neglected by real estate speculation. Corroborating this idea, Almeida (2010) describes that 25% of the world's population live in areas of natural hazards and that most of this population is found in developing countries due, above all, to the high level of vulnerability they face, since in these countries there are high levels of poverty, segregation/discrimination and a lack of political representation, which impede the development process. Intense urbanization in developing countries has occurred rapidly and with a lack of control on the part of public authorities in guiding the forms of occupation.

The frequent hypertrophy of Latin American cities is distinguished above all by their limited power to absorb the growing increase in population; due, of course, to the natural balance, but above all to an important rural exodus. This triggers the building of precarious and marginal housing concentrated in marginal neighborhoods. This situation occurs and develops on spaces estimated to be unfit for construction, either because they are unhealthy, dangerous or because of their open exposure to certain natural hazards (volcanic activity, landslides, flooding, etc.). They are a form of housing that concentrates an increasingly large population in Latin American cities (CHARDON, p.5; 1998).

Chambers (2006) points out that poverty makes individuals more vulnerable because it reduces their choice. This lack of choices is the motivation that leads socially segregated groups to settle in areas of high environmental vulnerability. It is also for this reason that countries with low social indicators record high rates of human loss when faced with potentially damaging natural phenomena, which in many cases are not necessarily extreme, as Chardon says (p. 6, 1998) *"Poverty increases vulnerability to risk and disasters increase poverty".*

It is notorious that many of the damaging effects of natural hazards in countries such as Brazil are the result of a lack of urban planning policies in cities, since most of the losses and damage are caused by events that are short-lived and concentrated in time and space, but which can be perceived in advance, since they are risks arising from the irregular occupation of inappropriate areas. This factor, coupled with the lack of choice for poor people to buy more environmentally stable areas in cities, results in the growing number of natural disasters and people affected - remembering that there is only risk if there is something or someone vulnerable.

Poverty is doubly a risk factor: it imposes the colonization of dangerous spaces and attracts the attention of people who don't have the resources to worry about environmental problems. Thus, the urban process is a creator and amplifier of risks (CHARDON, p.11, 1998).

From Robertier's (2012) perspective on the territorialization of risks in cities, it can be concluded that in these environments, which also have intense social dynamics, risks will be spatialized differently as a result of the actors who form them, concentrating on the areas of greatest vulnerability.

Today's society, especially in urban areas, is exposed to a range of risks, but the spatiality of these risks depends on the vulnerabilities that are so distinct (or not) in the urban environment. Zanirato et al. (2008), paraphrasing Beck, articulates:

If misery is hierarchical, pollution is democratic, which makes the risk society different from the class society, since its dangers are distributed in all directions and do not respect state borders, even though they affect social classes differently (ZANIRATO et al., p.5 2008).

The city combines a number of highly dynamic factors, both natural - given that it is inserted in a physical-natural substrate - and social, including local culture, which engender a variety of risks. However, social actors end up dictating who is most vulnerable in this environment.

The public authorities have a duty to establish the uses to which urban land is put.

of mechanisms such as the Master Plan, which guarantees the municipality's sustainable urban development policy, guiding public and private agents in the production and management of urban space. In this way, a city's Master Plan should aim to develop the social and environmental functions of the city as a whole, as well as of property, striving for a socially just use of urban land, ensuring the population's quality of life.

Thus, the Master Plan for the city of Natal, in Chapter III, Article 6, paragraph V, defines risk areas:

Areas likely to be affected by natural and/or man-made processes that cause adverse effects, located on and around slopes, areas susceptible to flooding and/or inundation, topographic shallows, railroad line right-of-way, high-voltage electricity transmission line right-of-way, areas over gas pipeline lines, dune flanks and adjacent areas, slopes and areas of instability subject to landslides and/or burial (Natal, 2007).

Social vulnerability is the consequence of a range of factors that permeate from individuals' living conditions to their perception of risks, making the attempt to measure and operationalize social vulnerability complex. However, some aspects can indicate not only the degree of poverty of a group, but above all access to facilities that guarantee quality of life and the possibility of social advancement.

In order to draw up an index of social vulnerability, not only indicators linked to income should be taken into account, as this is only one of the factors of vulnerability. Thus, issues linked to gender, age, family configuration, schooling and access to public infrastructure facilities will have a major influence on the different causes and consequences when in a situation of risk.

This work was based on the São Paulo Social Vulnerability Index, which aggregates multiple dimensions of poverty, including income indicators and other factors related to schooling and the family life cycle; moreover, this methodology allows the identification of areas according to the social vulnerability of the resident population, in other words, it enables the spatialization of social vulnerability.

The following chapter discusses how this methodology was applied to the reality of the capital city of Potiguar, pointing out which variables were used and how social vulnerability was spatialized in the city.

3. SOCIAL VULNERABILITY IN NATAL

Natal's historical factors and natural conditions have determined the pattern of urban land use and occupation in the city, as well as explaining the formation of socially segregated areas.

The dune fields, mangroves and riverbanks are occupied by a lower middle class population. Historically, according to Silva (2003), Natal had fishing villages such as Rocas, Areia Preta, Ponta Negra and Redinha; in addition to these, there were settlements in Passo da Pàtria and the Alecrim neighborhood that were occupied by a poorer population. Due to the great drought at the beginning of the 20th century, many farmers from the interior of the state sought refuge from the drought in the capital of Rio Grande do Norte and, for the most part, settled on the outskirts of the Ribeira and Cidade Alta neighborhoods in search of work or handouts. According to Silva (2003):

> Thus, the first peripheral neighborhoods emerged in the city. The largest of these was Alecrim, which to this day retains its popular appearance with its villas, small plots and high housing density. On the left bank of the Potengi River, other settlements developed, such as Igapó, adjacent to the railroad, built in 1916, and Redinha. This configuration remained until the early 1940s, when the advent of the Second World War changed the city, boosting its growth. Although in the center of the capital, the commercial houses were teeming with people and activity, in the more peripheral neighborhoods and villages (such as Alecrim, Passo da Patria and Rocas) the poor population increased, without having much opportunity for the best jobs (SILVA. 2003, p.73).

From the 1940s onwards there was a greater dynamism in the urbanization of Natal due to the arrival of the Americans in the state during the Second World War. The city's proximity to the European continent made it a strategic point for the Allied forces, and it became home to the North American air base. After this episode, the city ceased to have a hitherto inexpressive urban growth and began to attract new inhabitants, mainly from the interior of the state, who wanted to work on the new infrastructures for the reorganization of urban space that were being developed in the city (OLIVEIRA, 2003).

In the 1970s, investments were made at the federal level through the BNH (National Housing Bank), which promoted the implementation of housing programs in Natal, as had happened in other cities across the country. Through this initiative, two program models were created: the housing complexes promoted by COHAB (Companhia de Habitaçâo Popular), which was responsible for building homes for the low-income population, and the developments managed by INOCOOP (Instituto de Orientaçâo as Cooperativas Habitacionais do Rio Grande do Norte e Paraiba), which focused on building housing complexes on the city's most valuable land. The housing estates built in the city had up to 2,000 homes, and Silva (2003) points out that "without a plan or roadmap to organize this production, the housing estates were shaping the urban space of Natal, demarcating what would be the periphery and what would be the 'noble area'" (SILVA, 2003, p. 81).

The areas furthest away from the city center did not - and still do not - have urbanization to guarantee basic infrastructure for the population living there, and the geomorphological conditions were not conducive to occupation, so these areas became segregated and received the least socially and economically advantaged populations.

> These areas began to consolidate from the 1970s onwards, when Natal underwent accelerated development, registering only 30 agglomerations of subnormal housing; the low-income population or those in a recent state of urbanization began to occupy areas that were unsuitable from a geomorphological point of view or precarious because they had not yet been subjected to basic urbanization actions. These communities have gone through successive processes of "favelization", acquiring their own social dynamics, with a tendency towards exclusionary differentiation. (Secretaria Municipal de Meio Ambiente e Urbanismo de Natal, 2008; p.3).

The capital of Rio Grande do Norte has some natural barriers, such as the Potengi River, which divides the northern area from the other areas, and the dune system, which sometimes has a steep slope extending from the coast and into the city. This has meant that some areas of the city have been neglected by the occupation process due to the difficulty of accessing them, as well as the need to apply more elaborate engineering techniques.

> Between 1994 and 2002, urban expansion in Natal broke through natural boundaries which, in many places, made it difficult for the built environment to access them. During this period there was a major shift towards the northeast and south axes, and especially towards the west and southeast, due to the expansion of new neighborhoods such as Planalto (OLIVEIRA, 2003, p. 125).

Many areas of the Potiguar capital have a high degree of social and environmental vulnerability. In particular, these vulnerabilities overlap, especially in the city's peripheral regions.

The social vulnerabilities found in the poorest areas of the city are of various kinds, including not only issues linked to the lack of means of survival, such as income, housing and guaranteed food, but also social vulnerabilities linked to violence and drug use, which end up influencing the life expectancy of many young people, especially males.

However, this chapter focuses only on the measurement of social vulnerability based on variables from the 2010 IBGE Demographic Census. It is worth noting that the methodology used in this research is an adaptation of the São Paulo Social Vulnerability Index, prepared by the SEADE Foundation (Fundaçao Sistema Estadual de Analise de Dados) of the State of São Paulo (SEADE, 2008).

The importance of producing a social vulnerability index lies in the possibility of diagnosing which social groups are at risk, where these groups are located and the risk to which they are exposed. In this way, with spatialization, together with knowledge of the profile of those at risk and the type of risk they are exposed to, work aimed at prevention and mitigation will be more specific and will consequently be able to contribute more effectively to the promotion of public policies for risk and disaster management.

2.1. Methodology for drawing up Natal's Social Vulnerability Index

In order to analyze the socially vulnerable areas of the Potiguar capital, a social vulnerability index (SVI) was developed using socio-economic data obtained from the 2010 demographic census of the Brazilian Institute of Geography and Statistics (IBGE), in which the census sector is the smallest spatial unit of analysis. Thus, through the use of geographic information systems and geoprocessing programs, the information on the 895 census sectors[4] (figure 04)[5] could be spatialized.

4 The municipality of Natal has a total of 910 census sectors, however, 15 of them are considered special sectors, in this case, the special sector of barracks, military bases etc.; special sector of housing; special sector of asylums, orphanages, convents, hospitals etc. fall into this category.

5 The city's streets were added to the map of census tracts in order to distinguish census tracts with a low density of occupation.

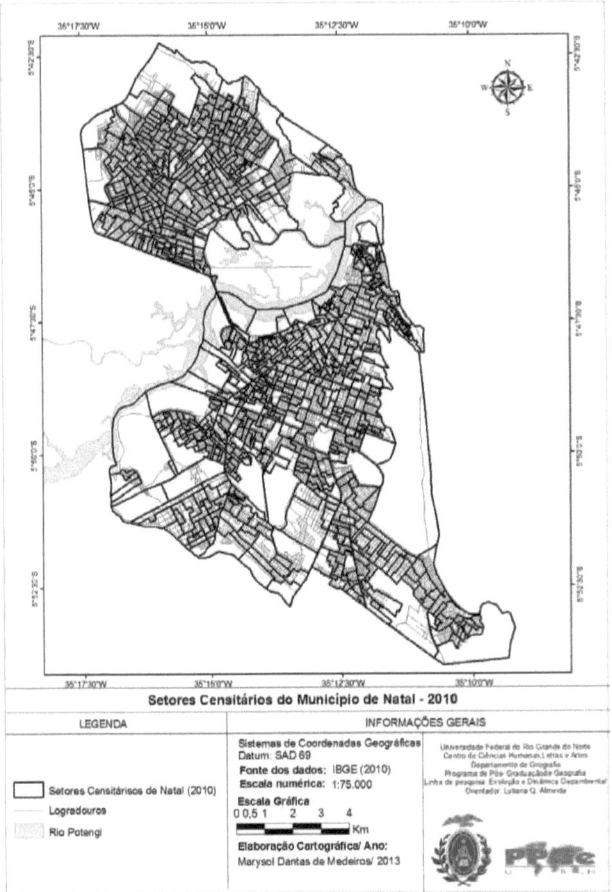

Figure 04: Census Sectors and Neighborhoods in the Municipality of Natal, 2010.**Source:** Prepared by Marysol Medeiros based on data from the 2010 IBGE Demographic Census.

The Social Vulnerability Index was created by combining 71 variables from the IBGE's 2010 Demographic Census that denote social and infrastructure disadvantages. Variables were selected that take into account income, gender, age, schooling, housing and street infrastructure. It was necessary to compile some variables, such as those related to age, because each age corresponded to a variable, so after putting them together, the result was 21 variables, as shown in the table below (Table 06).

Table 06: Variables from the IBGE Demographic Census (2010) used in the research

EVALUATION CRITERIA SOCIAL	VARIABLES

VULNERABILITY	
General Household Characteristics/ Concentration by Household/ Gender	V1-Households without a piped water supply.
	V2-Households with toilets without adequate sanitary sewerage
	V3-Homes without a bathroom
	V4-Households without garbage collection
	V5-Households without electricity or with illegal connections
	V6-Households with more than 6 residents
	V7-Households without a male resident
	V8-Woman responsible for household with 5 or more residents
	V9-Households in another condition (invaded land)
Gender / Age Group	V10-Female respondents
	VH-Responsible from 10 to 19 years old
	V12-Children from 0 to 9 years old
	V13-Improvised dwellings
Income	V14-Household with monthly per capita income of up to 2 minimum wages
	V15-Female caretaker with a monthly income of up to 2 minimum wages
Street infrastructure	V16-Dwellings in a street without public lighting
	V17-Dwellings in unpaved areas
	V18-Dwellings in areas without culverts or manholes
	V19-Dwellings in areas with open sewers
	V20-Households in areas with
Illiteracy	V21-Alphabets by census tract

Source: Marysol Medeiros based on data from the 2010 IBGE Demographic Census.

The preparation of the Social Vulnerability Index involved the use of multivariate analysis, which is the grouping together of various variables. This compilation resulted in the reduction of the twenty-one variables into just six factors, this technique being called Factor Analysis.

Factor analysis makes it possible to create a matrix of correlations or covariances between the selected variables, based on their dependence. This allows the number of variables to be reduced, with the aim of condensing the information contained in the various original variables into a smaller set of new composite dimensions, better known as factors, with minimal loss of information.

Factor analysis was used to obtain factor estimates for each sector. As Table 01 shows, 69.30% of the total variability of the 21 variables is explained by six factors. Further analysis was carried out using these six factors.

Table 01: Total variance explained

Factor	**Initial eigenvalues**		
	% of variance	% cumulative	Total
1	7,373	35,110	35,110
2	2,156	10,267	45,378
3	1,659	7,900	53,278
4	1,387	6,605	59,883
5	1,011	4,812	64,695
6	0,969	4,613	69,308

Source: Research data

With this factor analysis, each factor is made up of a set of variables, in which the variables with the highest loadings (values in bold) are considered to be the most important and have the greatest influence on the name or label selected for the factor, as shown in Table 02:

Table 02: Component factor analysis matrix: complete and reduced sets of variables

SET	COMPONENTS					
COMPLETE WITH VARIABLES	**1**	**2**	**3**	**4**	**5**	**6**
Households without a piped water supply.	**0,569**	-0,112	-0,077	-0,250	0,153	0,079
Households with toilets without sanitary sewerage	0,286	**0,643**	0,043	-0,025	0,092	0,018
Households without a bathroom	0,136	0,062	**0,833**	-0,013	0,065	-0,050
Households without garbage collection	0,154	0,175	**0,714**	-0,062	0,345	0,031
Households without electricity or connections	0,389	0,258	**0,517**	-0,122	-0,026	0,166

Households with more than 6 residents	**0,742**	0,257	0,211	0,205	-0,038	0,133
Households without a male resident	-0,020	-0,159	-0,035	**0,876**	0,010	0,116
Woman responsible for household with 5 or more residents	**0,846**	0,092	0,132	0,176	-0,019	0,063
Households in another condition (invaded land)	-0,005	-0,050	**0,701**	0,153	-0,205	-0,014
Female managers	**0,525**	0,072	0,022	**0,797**	0,052	-0,001
10 to 19 year olds in charge	**0,469**	0,123	0,189	0,258	0,203	-0,192
Children from 0 to 9 years old	**0,617**	**0,600**	0,302	0,227	0,051	0,015
Makeshift homes	0,107	0,044	0,019	0,079	0,009	**0,952**
Household with monthly per capita income of up to 2 minimum wages	**0,667**	**0,561**	0,123	0,289	-0,045	0,044
Responsible female with a monthly income of up to 2 minimum salaries	**0,783**	0,345	0,073	0,348	0,000	-0,025
Households in areas without public lighting	0,088	0,214	0,019	0,083	**0,825**	0,000
Households in unpaved areas	0,037	**0,717**	0,232	0,104	0,233	-0,011
Households in public places no manholes	0,234	**0,403**	0,052	**0,650**	0,082	-0,039
Households in areas with open sewers	0,323	**0,655**	-0,051	0,010	-0,326	-0,063
Households in areas where garbage accumulates	-0,009	**0,685**	0,007	-0,007	0,158	0,063
Illiterates by census tract	**0,617**	**0,487**	0,384	-0,018	0,044	0,021

Source: Research data.

As a result, Factor 01 was named "General Characteristics of Residents" because it has higher values for variables that correspond mostly to the profile of the residents. Factor 02 has higher weights in variables linked to the conditions of the neighborhoods, and is labeled "Characteristics of the Surroundings of the Households". Factor 03 covers the variables that denote the conditions of the homes, and is therefore called "Household Infrastructure". Factor 04 brings with it the variables linked to gender and the urban infrastructure that corresponds to rainwater runoff, making this factor be called

"Household without a male resident and surroundings without a rainwater runoff system". Factor 05, on the other hand, had a high weighting only in the area related to public lighting, and was therefore called "Households in areas without public lighting". Factor 06 refers to dwellings that were not intended for habitation but are used for this purpose, i.e. they are improvised, which is why this factor was called "Improvised Dwellings". The variables that make up each factor are organized as follows

in table 07:

Table 07: Variables that make up each factor of the Social Vulnerability Index

FACTORS	VARIABLES
Residents' general characteristics	Woman responsible for household with 5 or more residents
	Female manager with a monthly income of up to 2 minimum salaries
	Households with more than 6 residents
	Household with monthly per capita income of up to 2 minimum wages
	Children from 0 to 9 years old
	Illiterates by census tract
	Households without water supply
	Female managers
	Responsible persons aged 10 to 19
Characteristics of the homes' surroundings	Households in areas without
	Households in areas with an accumulation of
	Households in areas with open sewers
	Households with toilets without adequate sanitary sewerage
	Children from 0 to 9 years old
	Household with monthly per capita income of up to 2 minimum wages
	Illiterates by census tract
	Households in areas without
Household infrastructure	Households without a toilet
	Households without garbage collection
	Households in another condition (invaded land)
	Households without electricity or with illegal

	connections
Household with no male resident and surroundings with no rainwater drainage system	Households without a male resident
	Female managers
	Households in areas without manholes
Households in areas without public lighting	Households in areas without public lighting
Households	Makeshift homes

Source: Research data.

Using the *ArcGIS 9.3* geoprocessing program, it was possible to spatialize these factors. However, in order to form homogeneous groups of sectors, the *Natural Breaks* statistical method used in the program was used, in which the data was divided into six classes ranging from very low social vulnerability, covering the sectors with the best social indicators, to very high social vulnerability, representing the census sectors with the worst social indicators (see figure 06).

By analyzing each factor separately, it is possible to identify the main problems affecting the population of each census tract. In this way, public policies become more targeted and perhaps more effective.

Analyzing the factors, it can be seen that each one mixes variables with characteristics intrinsic to the individual, such as: gender, income, schooling, age; as well as attributes linked to the infrastructure of the homes and the streets in which they are located.

Evaluating each factor separately, it is possible to draw the following conclusions: the first factor, "Residents' general characteristics", which includes variables linked to gender, income, age and education, has a concentration of sectors with social vulnerability between very high and high in neighborhoods on the west side, mainly Guarapes, Planalto, Cidade Nova, Quintas and Bom Pastor; and on the north side, in the neighborhood of Redinha, more specifically in the community of Africa and in the neighborhood of Nossa

Senhora da Apresentaçâo. The eastern zone has some sectors in the neighborhoods of Mae Luiza and Rocas (in the Sao José do Jacó community). And the southern zone has small sectors in the neighborhoods of Nova Descoberta, Neópolis and Ponta Negra.

Factor 02, "Characteristics of the surroundings of the homes", brings together variables that concern urban infrastructure and public services provided to the neighborhoods, such as drainage equipment, sanitary sewage, paving and garbage collection. Taking these variables into account, a large part of the northern zone, especially the neighborhoods of Redinha, Pajuçara, Lagoa Azul and Nossa Senhora da Apresentaçao, have several sectors with very high or high social vulnerability for this factor. In the west, the Planalto and Cidade Nova neighborhoods stand out. The other areas have very low and medium to low values.

Factor 03, "Household infrastructure", includes variables such as households without bathrooms, without electricity, without garbage collection and on "invaded" land. In this regard, only one sector in the Guarapes neighborhood, two in the Planalto neighborhood, one in Neópolis and four in Nossa Senhora da Apresentaçao had very high or high social vulnerability. The other sectors of the city showed low levels of vulnerability.

Factor 04, "Households with no male residents and surroundings with no rainwater drainage system", is considered the most controversial factor, since sectors located in prime areas of the city were given high vulnerability indices. This is because in the so-called "noble" areas of the city there is no drainage project, and many streets are not even paved. Not only that, but in these areas there is a higher concentration of women who are responsible for the household or who simply don't have a male presence in their homes.

The factor "Households in areas without public lighting", factor 05, showed a vulnerability value equivalent to "Medium to Low" in most of the city, with some specific sectors in Candelaria, Nova Descoberta and Ribeira without this service. The northern zone, on the other hand, had more sectors with very high and high vulnerability, especially in the Nossa Senhora da Apresentaçao neighborhood.

Factor 06, "Improvised dwellings", refers to sectors with buildings that are not intended for habitation, but which are used for this purpose. It can be seen that the more commercial the neighborhood, the more sectors will be in this condition, so the neighborhoods of Alecrim, Capim Macio, Neópolis, Pajuçara and Nossa Senhora da Apresentaçao have sectors with these characteristics.

At first glance, the maps seem to present a certain inconsistency in the information, as well-known areas of the Potiguar capital with high income standards are marked with social vulnerability values ranging from "Medium to High" to "Very High". However, it should be pointed out that social vulnerability is not just about measuring the characteristics directly linked to individuals, nor should it be based solely on residents' income. Social vulnerability encompasses a universe of nuances that are not always covered due to a lack of data. In this way, the qualities of the infrastructure in which individuals live are also considered to be of fundamental importance when measuring social vulnerability, given that poor urban facilities can put people from higher social classes at risk. Although the resilience capacity of social classes with higher incomes is higher, exposure to danger is also high in this case.

By overlaying the six factors, it was possible to map the census tracts with the highest social vulnerability indices in the city of Natal. Using the same statistical method used in the Arcgis 9.3 geoprocessing program to draw up the maps for each factor in isolation, the following result was obtained, as can be seen in figure 05.

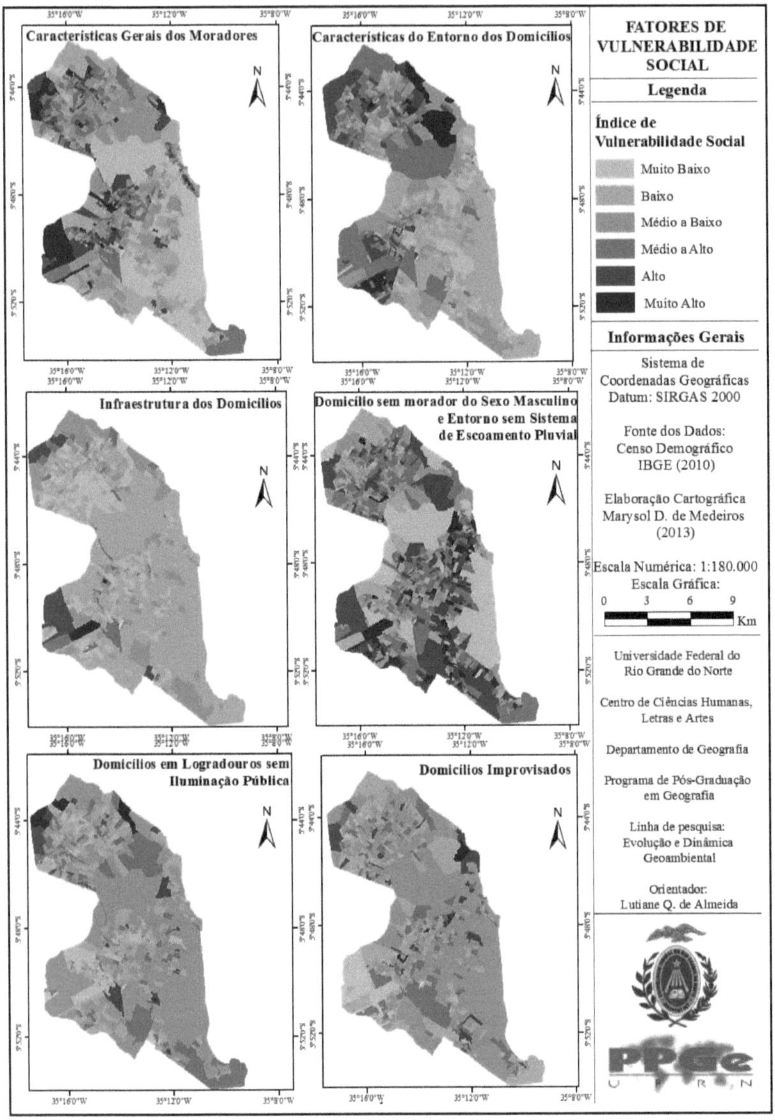

Figure05: SVI factors. **Source:** Prepared by Marysol Medeiros based on data from the 2010 IBGE Demographic Census.

Even though it appears to be a generalized result, the spatialization of these overlapping factors provides a diagnosis of the city's social vulnerability conditions, making it possible to refine the analysis by mapping the factors or variables in isolation. Analyzing the data and with the help of the following graph (graph 01), the city of Natal has few census tracts classified as being of very high social vulnerability, totaling only 11 tracts (1%) in the whole city, of which 01 tract is in the east, in the Alecrim district, with 1001 people and 325 households; 08 in the north, 04 of which are in the Nossa Senhora da Apresentaçâo district with 6.175 residents and 1,719 households, 02 in Redinha with 2,911 residents and 819 households, 01 in Pajuçara with 1,555 residents and 449 households and 01 in Lagoa Azul with 981 residents and 283 households; 02 in the west, one in the Cidade Nova neighborhood with 1,319 residents and 392 households and the Guarapes neighborhood with 2,483 residents and 688 households. No sectors with very high social vulnerability were registered in the south and the total area of census sectors with very high social vulnerability is 417,469m .[2]

42 census tracts (5%) were classified as having high social vulnerability, of which 01 is located in the east (Alecrim neighborhood); 17 tracts in the north (03 in the Igapó neighborhood, 06 in Lagoa Azul, 04 in Pajuçara, 03 in Redinha and 01 in Nossa Senhora da Apresentaçâo); 19 sectors in the west (02 in Bom Pastor, 08 in Planalto, 02 in Guarapes, 04 in Felipe Camarâo and 02 in Quintas) and 05 sectors in the south (01 in Nova Descoberta, 02 in Neópolis and 02 in Candelària).

With medium to high social vulnerability, 18% of all census tracts are covered, distributed as follows: 20 tracts in the east, 82 in the north, 36 in the west and 28 in the south.

Figure 06: Spatialization of the Social Vulnerability Index. **Source:** Prepared by Marysol Medeiros based on research data.

As can be seen in the graph below, 24% of the city's sectors have a degree of social vulnerability between medium high and very high. On the other hand, 40% of the sectors are in a situation of low social vulnerability, in addition to the 36% that have medium-low social vulnerability.

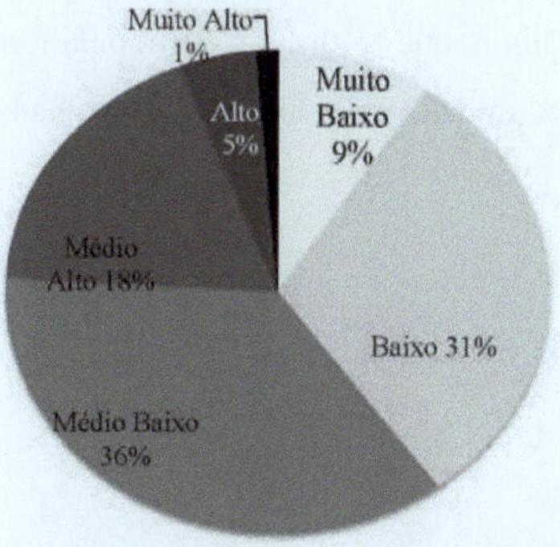

Graph 01: Distribution of census sectors according to degree of vulnerability **Source:** Prepared by Marysol Medeiros based on research data.

However, the 11 census tracts with very high social vulnerability encompass 2% of Natal's population (see graph 02), i.e. approximately 16,425 inhabitants face adversities linked to lack of income, education, urban infrastructure or precarious housing. To this contingent must be added the 7% of the population with high social vulnerability, i.e. approximately 55,630 inhabitants, giving a total of close to 72,055 inhabitants with high social vulnerability.

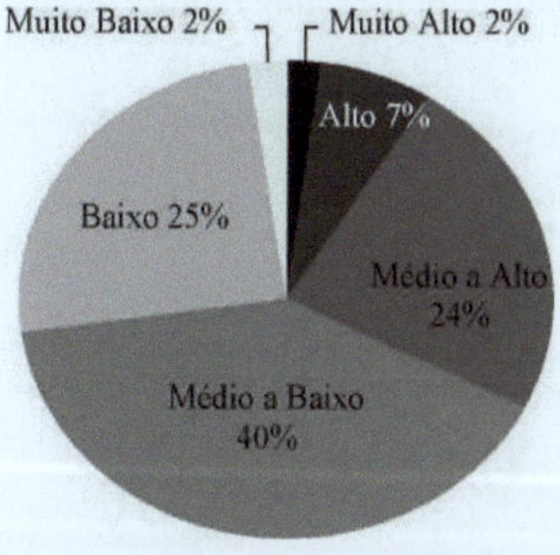

Graph 02: Distribution of the population of Natal according to degree of vulnerability. **Source:** Prepared by Marysol Medeiros based on survey data.

Analyzing the percentage distribution of census tracts based on social vulnerability classes by administrative region of the city, as well as the number of inhabitants in each vulnerability group, it can be seen that the east zone has the most tracts with very low social vulnerability (45%), followed by the west zone with 30% of tracts, the south zone with 19% and the north zone, which has only 6% of tracts in this class. However, when analyzing the population graph, the west zone has a greater number of inhabitants despite having fewer sectors than the east zone. This is due to the higher population density in these sectors of the west zone, especially those with older urbanization, which are very densely populated.

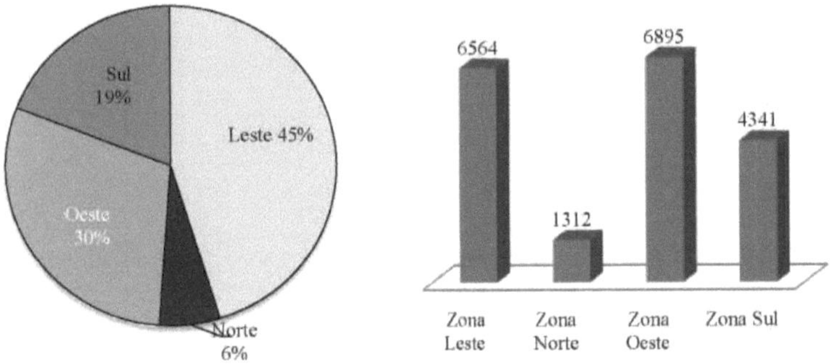

Graph 03 (Left): Percentage of sectors with very low social vulnerability by administrative region. **Right:** Population by administrative area with very low social vulnerability. **Source:** Prepared by Marysol Medeiros based on survey data.

For the low social vulnerability class, there is a balance in the distribution of both sectors and population between the city's four administrative zones, with the southern zone responsible for the largest number of sectors and population for this vulnerability group.

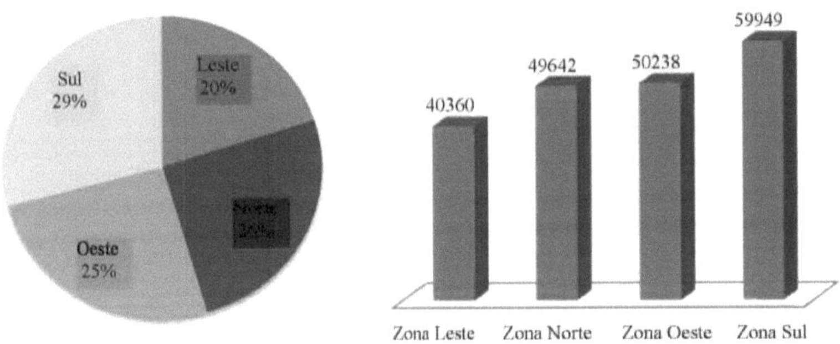

Graph 04 (Left): Percentage of sectors with low social vulnerability by administrative region. **Right:** Population per administrative area with low social vulnerability. **Source:** Prepared by Marysol Medeiros based on survey data.

The changes in the graphs with regard to the sectors in the northern part of the city are evident, since the higher the degree of social vulnerability the percentage of sectors increases, so that for the medium to low social vulnerability class the northern part of the city begins to show itself to be more significant than the other regions of the city.

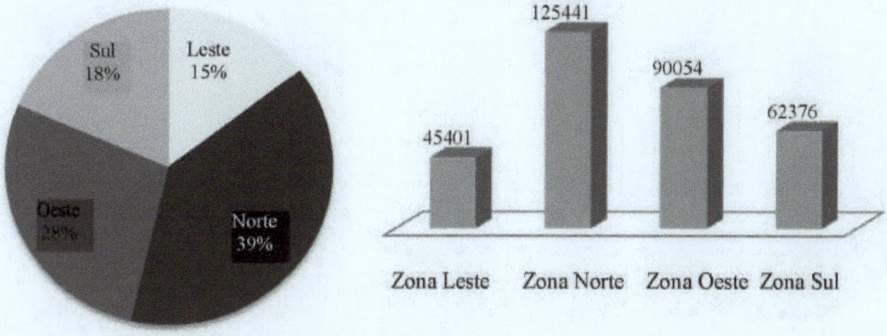

Graph 05 (Left): Percentage of sectors with medium to low social vulnerability by administrative region. **Right:** Population by administrative area with medium to low social vulnerability. **Source:** Prepared by Marysol Medeiros based on survey data.

As can be seen in the following graph, almost half of the census sectors with medium to high vulnerability (49%) are in the north of the city, i.e. almost 93,000 people. It is worth remembering that only 24% of Natal's population is in medium to high vulnerability.

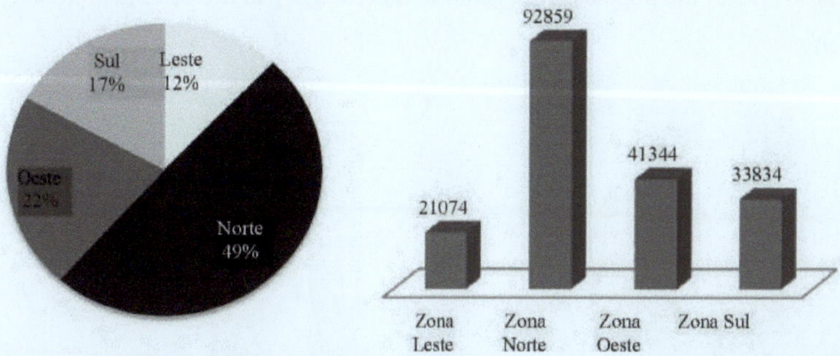

Graph 06 (Left): Percentage of sectors with medium to high social vulnerability by administrative region. **Right:** Population by administrative area with medium to high social vulnerability. **Source:** Prepared by Marysol Medeiros based on survey data.

With regard to the census sectors with high social vulnerability, 45% are in the west and 41% in the north of the city. Adding up the population with high social vulnerability in these two regions gives a total of 48,739 people. Remembering that there are around 55,630 people in high social vulnerability in the city as a whole, the west and north zones alone account for almost 88% of the population in high social vulnerability in Natal.

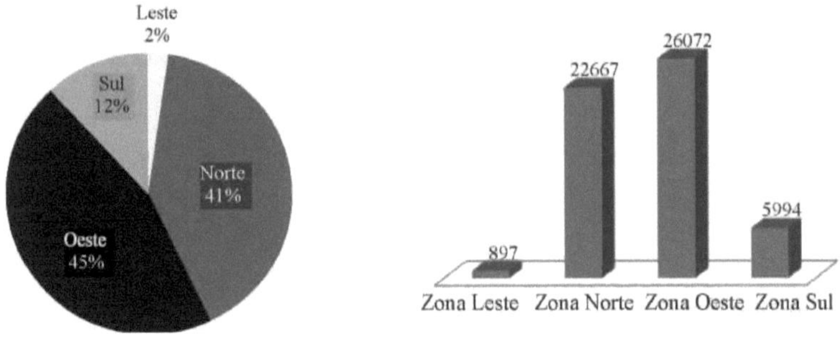

Graph 07 (Left): Percentage of sectors with high social vulnerability by administrative region. **Right:** Population by administrative area with high social vulnerability. **Source:**

Prepared by Marysol Medeiros based on survey data.

When it comes to very high social vulnerability, the precarious social and infrastructural conditions found in the north of the city are evident, since of the 16,425 inhabitants with very high social vulnerability, 11,622 are in the north, making 73% of the census sectors and almost 71% of the population of Natal in this category in this administrative region of the city.

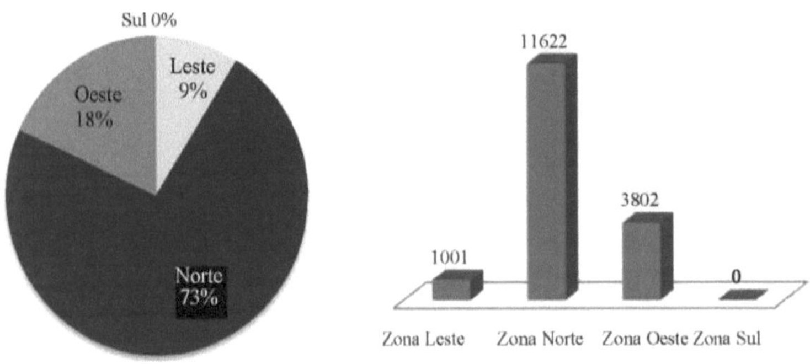

Graph 08 (Left): Percentage of sectors with very high social vulnerability by administrative region. **Right:** Population by administrative area with very high social vulnerability. **Source:** Prepared by Marysol Medeiros based on survey data.

Each social vulnerability class corresponds to a group, so group 01 is made up of census tracts with very high social vulnerability, group 02 includes census tracts with high vulnerability, group 03 with medium to high vulnerability, group 04 with medium to low social vulnerability, group 05 includes tracts with low social vulnerability and group 06 includes tracts with very low social vulnerability.

Analyzing each group separately, we can see some patterns of vulnerability and the profile of the population that makes up each one:

- *Group 01 - Very High Social Vulnerability*

This group includes the census tracts with the highest values in the social vulnerability index, due to the low economic conditions of residents, precarious housing and lack of public services and urban infrastructure.

As can be seen in table 03, this group has 11 sectors (1% of the total) and is home to a population of 16,425 inhabitants (2%) in very high social vulnerability, this population occupying 4,675 households.

The determining factors for the high social vulnerability of this group were the general characteristics of the residents, the characteristics of the surroundings of the homes and the infrastructure of the homes. Thus, the social profile of the residents of these sectors is made up of a low-income contingent, a high number of illiterate people, a high number of children and women; the homes have a high demographic density, the presence of homes with no water supply, no toilets or toilets without adequate sanitation; most of the streets are not paved, there is no garbage collection, so there is an accumulation of solid waste in the streets, the presence of wastewater effluents flowing superficially into the streets, there is no drainage system and sometimes the houses are built on invaded land, so that the residents can be evicted at any time. Most of these areas are located on the edge of the city, with several empty portions and incipient urbanization, with the exception of the census tract located in the Alecrim neighbourhood.

- *Group 02 - High Social Vulnerability*

Group 02 covers the census sectors with high social vulnerability, with 42 sectors (5% of the total) with a population of 55,630 inhabitants (7%) occupying 16,2345 homes. The determining factors for the high social vulnerability of this group were: the surroundings of the homes (mainly in the sectors of the Planalto, Redinha and Pajuçara neighborhoods), with streets in similar conditions to those found in the aforementioned group and the characteristics of the residents, which weighed heavily on the high social vulnerability of sectors in the Planalto, Redinha and Nossa Senhora da Apresentaçâ neighborhoods. In the neighborhoods located in the south, such as Neópolis and Candelària, the factor of households with no male residents and surroundings with no rainwater drainage system was more relevant in classifying these areas as highly socially vulnerable. Generally speaking, despite the precariousness of some aspects, these sectors have better household infrastructure than the sectors in group 01. Most of the census tracts in this group are located in the west and north of the city.

- *Group 03 - Medium to High Social Vulnerability*

Group 03 brings together the census sectors with medium to high social vulnerability, with 166 sectors (18% of the total) with a population of 189,111 inhabitants (24%) occupying 55,573 households.

The preponderant factor for the degree of social vulnerability of this group is linked to households with no male residents and surroundings with no rainwater drainage system. To a lesser extent, the lack of public lighting had an influence on the classification of these areas.

The census tracts classified as having medium to high social vulnerability are well distributed throughout the city, but with a greater concentration in the south and some neighborhoods in the north, such as Igapó.

- *Group 04 - Medium to Low Social Vulnerability*

Group 04 is made up of census tracts with medium to low social vulnerability, with 328 tracts (36% of the total) and a population of 323,272 inhabitants (40%) occupying 94,958 households. It is the largest group, both in terms of the number of sectors, population and territorial extension.

Group 04 is made up of census tracts located in the most central parts of the city. Many census tracts have consolidated urban infrastructure, however, they are home to historically socially segregated communities, such as the Mâe Luiza neighborhood, the Sâo José do Jacó community in the Rocas neighborhood and the Passo da Pàtria community in the Cidade Alta neighborhood. This is because these sectors, despite the economic precariousness of their residents, have some urban facilities, such as food, sanitation and drainage systems.

- *Group 05 - Low Social Vulnerability*

Group 05 is made up of census tracts with low social vulnerability. It is the second largest group, with 279 tracts (31% of the total), a population of 200,189 inhabitants

(25%) and 59,749 households.

It is made up of census tracts located in the central areas of the city and in the so-called noble neighborhoods, mainly in the east, such as the Tirol, Lagoa Seca, Cidade Alta and Areia Preta neighborhoods, and in the Pitimbu, Capim Macio and Ponta Negra neighborhoods in the south, with some tracts located in the west and north. These are areas of the city that have regular urban infrastructure and services, although some areas do not have a rainwater drainage system.

- *Group 06 - Very Low Social Vulnerability*

Group 06 includes the census tracts classified as having very low social vulnerability, with only 84 census tracts (9%) and an estimated population of 19,112 inhabitants. It is the second smallest group in terms of number of people.

Its census sectors have some peculiarities, as they are not necessarily those with the best economic and social conditions in the city. In fact, sectors located in Capim Macio, near the Campus of the Federal University of Rio Grande do Norte, sectors in the neighborhoods of Lagoa NoVa, Lagoa Seca, Tirol, Ponta Negra and Petrópolis have residents with high economic standards. However, there are sectors such as the one that encompasses the City Park and the mangroves of Environmental Protection Zone 08 that have no occupation and have been classified as having very low social vulnerability. Another relevant point is the military villages, which were also classified as having very low social vulnerability.

The table below summarizes the main characteristics of each social vulnerability group.

Chart 08: Characteristics of social vulnerability groups

Group	Residents' characteristics	Household characteristics	Street Characteristics	Social Vulnerability
1	• Mostly children and young people; • Women and young people as heads of household; • Low-income population; • High number of	• High population density; • No water supply; • No toilets or toilets without proper sanitation; • No garbage collection; • Improvised homes or homes built on invaded land; • No electricity;	• No pavement; • Garbage accumulation; • No rainwater drainage system; • Open sewage; • Without public lighting.	Very high

	illiterate people.	• Without the presence of men.		
2	• High numbers of women and young people as heads of household; • Low-income population.	• High population density; • With water supply; • Toilets with adequate sanitation; • No garbage collection; • With electricity; • Without the presence of men.	• No pavement; • Garbage accumulation; • No rainwater drainage system; • Open sewage; • Without public lighting.	High
3	- High number of women as heads of household.	• With water supply; • Toilets with adequate sanitation; • With garbage collection; • With electricity; • Without the presence of men.	• With paving; • No garbage pile; • No rainwater drainage system; • No open sewers; • With public lighting.	Medium to high
4	• Mostly children and young people; • Women and young people as heads of household; • Low-income population; • High number of illiterate people.	• High population density; • With water supply; • With toilets or toilets without proper sanitation; • With garbage collection; • With electricity;	• With paving; • With a pile of garbage; • With rainwater drainage system; • With open sewage; • Without public lighting.	Medium to Low
5	• High income; • High number of literate people; • Mostly adults.	• Low population density; • With water supply; • Toilets with proper sanitation; • With garbage collection; • With electricity;	• With paving; • No garbage pile; • No rainwater drainage system; • No open sewers; • With public lighting.	Low
6	• High income; • High number of literate people; • Mostly adults.	• Low population density; • With water supply; • Toilets with proper sanitation; • With garbage collection; • With electricity;	• With paving; • No garbage pile; • With rainwater drainage system; • No open sewers; • With public lighting.	Very low

Source: Research data; data adapted from the 2010 IBGE Demographic Census.

Table 03 below summarizes the amount of population per census tract, as well as the distribution in each group.

Table 03: Correlation between the Social Vulnerability Index groups and the numbers of the corresponding census tracts and their respective populations

Group	Sectors Census		Population		Social Vulnerability Index
	Abs.	**%**	**Abs.**	**%**	
1	11	1	16.425	2	Very High Social Vulnerability
2	42	5	55.630	7	High Social Vulnerability
3	166	18	189.111	24	Medium to High Social Vulnerability
4	328	36	323.272	40	Medium to Low Social Vulnerability
5	279	31	200.189	25	Low Social Vulnerability
6	84	9	19.112	2	Very Low Social Vulnerability
Total	910	100%	803.739	100%	

Source: Research data; data adapted from the 2010 IBGE Demographic Census.

3.2Spatialization of Natal's Social Vulnerability Index

Natal has 910 census tracts, of which 164 (18%) are in the East Zone, 311 (34%) in the North Zone, 244 (27%) in the West Zone and 191 (21%) in the South Zone.

To facilitate the diagnosis of the areas of Social Vulnerability in the city of Natal, the analysis will be based on the areas divided into the four Regions

administrative offices in the capital.

3.2.1 South Zone

Natal's South Zone is made up of eight neighborhoods (figure 07): Ponta Negra, Neópolis, Pitimbu, Capim Macio, Candelària, Parque das Dunas, Nova Descoberta and Lagoa Nova which bring together various census tracts containing a population of a high economic standard.

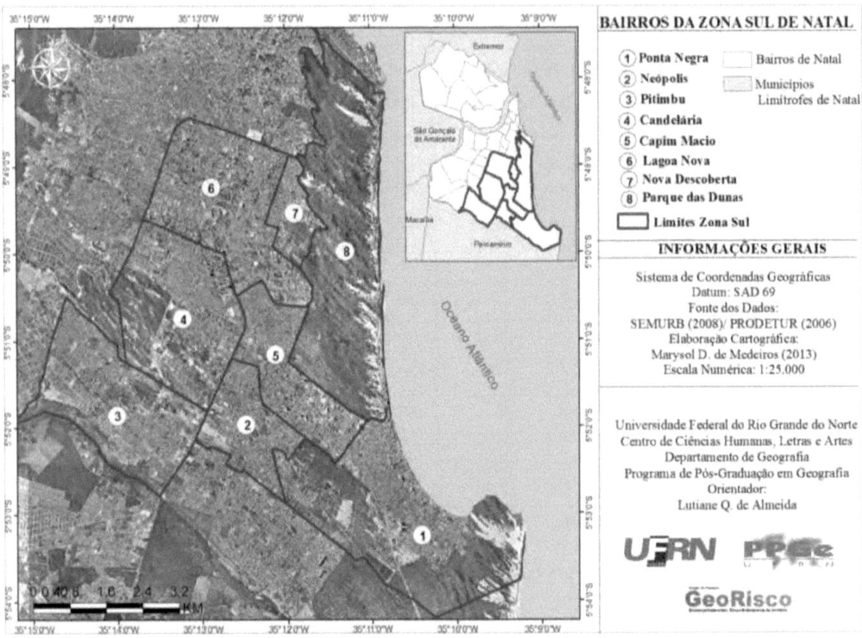

Figure 07: Neighborhoods in the Southern Administrative Region of Natal, RN. **Source:** Prepared by Marysol Medeiros using data from PRODETUR (2006) and SEMURB (2008).

However, Vulnerability is not only made up of the economic conditions of individuals; in this way, the South Zone has several census sectors with social vulnerability that are related to the lack of urban infrastructure, such as the absence of paving and drainage, variables that are directly linked to the risk of flooding.

The Capim Macio neighborhood, considered to be upper-middle class, is vulnerable due to the fragility of its roads in terms of drainage conditions. Some of the neighborhood's streets have been sealed off and many others remain unpaved, so, with exposed and compacted soil, it is difficult for water to percolate. In addition, the paved streets do not have a drainage system to prevent water from concentrating in the areas with the lowest topography.

The result of this combination of characteristics is that, during periods of heavy rainfall, these roads become veritable ponds, causing a great deal of inconvenience to residents, such as difficulties in getting around and material losses.

Another factor to consider in these neighborhoods is the history of urban occupation, where densification has taken place between the interdunal cordons present in this region of the city. As a result, several temporary lagoons resurface when the level of the underground aquifer rises and overflow, affecting neighboring areas. Although there has been a lot of investment in the construction of rainwater harvesting ponds, flooding is recurrent.

The Neópolis neighborhood, for example, has several natural lagoons which, during periods of heavy rainfall, overflow and hit the homes and buildings around them. In addition to the lack of urban infrastructure in terms of the drainage system, the sectors in this neighborhood with high social vulnerability indices also have a number of households where only women live, and the literature on studies of social vulnerability proves that women are more vulnerable than men. Thus, gender, together with the precarious drainage system, played a major role in the high levels of social vulnerability in these southern neighborhoods.

Other points of social vulnerability found in the South Zone are in the Candelària neighborhood, where you can find a community located on Travessa Senador Salgado Filho, also known as Viaduto - due to its proximity to the Ponta Negra Viaduct. This community has been established for over thirty years.

The houses are more modest than those in the aforementioned neighborhoods; they have no sidewalks and the main entrance faces directly onto the street, thus increasing the risk of undrained rainwater entering the homes; with unfinished sidewalks and narrow alleys, there is a concentration of surface rainwater and the discharge of wastewater onto the surface; and the drainage system is obstructed by garbage (figure 08). It is a densely occupied area with basic, albeit precarious, infrastructure. Close to the community is the catchment pond for the Avenida da Integraçâo, which constantly overflows.

Figure 08: Images of the Viaduto/Candelaria community. Top left: alley in the community showing the lack of planning in the occupation of the area. Top right: culvert in the Viaduto community obstructed by a build-up of garbage. Bottom left: catchment pond on Avenida da Integraçâo about to overflow and full of garbage. Bottom right: entrance to the Viaduto community with a build-up of water. **Source:** author's personal archive, August 2013.

Next to the Viaduto community, in the Pitimbu neighbourhood, more precisely in San Vale, an area of the neighbourhood with homes of a high economic standard, there is a sector characterized by not having any urban infrastructure. The buildings in this sector are predominantly flats and commercial establishments. On the banks of Avenida dos Xavantes, near Rua da Sorveria, there is a makeshift catchment pond that constantly fills up and overflows, making it impossible for vehicles of any size to pass along the avenue, as well as hitting the surrounding buildings with a water slide that easily reaches more than a meter in height. The following images (figure 09) illustrate the conditions found in the sectors described above.

Figure 09: Images of the sector in San Vale/Pitimbu. Top left: mark that the water has reached on the wall of the establishment, approximately one meter high, next to the improvised catchment pond on Avenida dos Xavantes. Top right: extension of the water overflow from the catchment pond on Avenida dos Xavantes, preventing even large vehicles from passing. Bottom left: houses on the banks of the lagoon being invaded by muddy water. Bottom right: houses on the other side of Avenida dos Xavantes with sidewalks a meter and a half high, completely covered by the water. **Source:** author's personal archive, September 2013.

The Nova Descoberta neighborhood has some sectors with high levels of social vulnerability, especially in relation to factors linked to the conditions of the residents, since it has households with low-income families, a high density of people per household, women as heads of household, as well as issues linked to the lack of adequate sanitary sewage, with wastewater flowing superficially through the streets, along with a large amount of accumulated garbage.

In the same conditions is the sector that covers the Coqueiros community, in the Lagoa Nova district, which is an area of high social vulnerability, with accumulations of garbage and waste water in the streets, as well as leaks in the water supply pipes. It is a densely occupied area with basic infrastructure, but in a precarious state. It is an old community with urbanization problems, as shown in figure 10.

Figure 10: Images of the Coqueiros/Lagoa Nova community. Left: Accumulation of garbage in the streets, since urban waste collection does not enter some streets because they have no exit. Right: Waste water thrown out into the open, a source of contamination. **Source:** author's personal archive, May 2013.

Also in the Lagoa Nova neighborhood are the crosswalks of Rua Tabeliâo Manoel Procópio and Rua Aluizio Bezerra. These are areas of medium to high social vulnerability due to the general conditions of the residents, but above all due to the lack of rainwater drainage. Both have the same urban characteristics: they are old settlements with modest masonry houses; paved streets with the presence of culverts and manholes on the main roads that surround them, but they are obstructed due to the accumulation of garbage; with narrow, fully cemented alleys and irregular sidewalks, but in most cases there are none, so that the main entrance to the houses faces directly onto the street, thus facilitating the entry of water; when it rains, since they are constant flooding points. The images below illustrate the situations described above (Figures 11 and 12).

Figure 11: Images of sectors in the Lagoa Nova neighborhood. Left: houses without sidewalks and with protective walls at the entrance to contain flooding. Right: manhole blocked by garbage. **Source:** author's personal archive, May 2013.

Figure 12: Images of Rua Aluizio Bezerra/ Lagoa Nova. Left: accumulation of garbage in the street. Right: narrow alley completely waterproofed due to soil cementation, making it difficult for water to percolate. **Source:** author's personal archive, September 2013.

Of the 191 census tracts that make up the southern zone, 16 tracts (8%) have a very low social vulnerability index; 81 tracts (42%) have low social vulnerability; 61 tracts (32%) are classified as having medium to low social vulnerability; 28 tracts (15%) have medium to high vulnerability; only 05 tracts (3%) have high social vulnerability and no tracts have been registered as having very high social vulnerability. This data can be seen in the following graph (Graph 09):

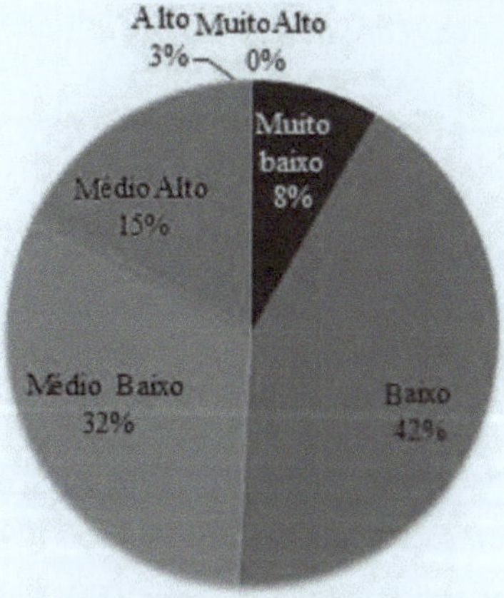

Graph 09: Graph of the distribution of census tracts based on the degree of social vulnerability for the southern zone. **Source:** Prepared by Marysol Medeiros based on survey data.

3.2.2 East Zone

The East Zone of Natal has twelve neighborhoods (figure 13): Tirol, Lagoa Seca, Alecrim, Barro Vermelho, Cidade Alta, Petrópolis, Ribeira, Rocas, Santos Reis, Praia do Meio, Areia Preta and Mae Luiza. This region of the city has profound social inequalities, with sectors with households of a very high economic standard alongside sectors of high social vulnerability.

The eastern zone is home to well-known areas of social vulnerability in the city, such as the Sao José do Jacó community, located in the Rocas neighborhood; the Passo da Patria community in the Cidade Alta neighborhood and the Barro Duro, Sopapo and Alto da Colina communities in the Mae Luiza neighborhood, and Aparecida between the Mae Luiza and Areia Preta neighborhoods.

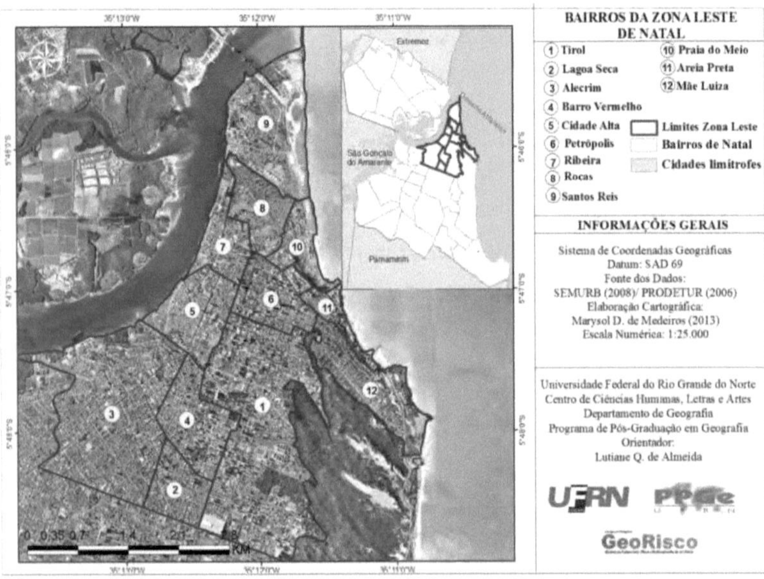

Figure 13: Neighborhoods in the Eastern Administrative Region of Natal, RN. **Source:** Prepared by Marysol Medeiros using data from PRODETUR (2006) and SEMURB (2008).

The eastern zone is made up of 164 census tracts, of which 38 tracts (23%) have very low social vulnerability; 56 tracts (34%) have low social vulnerability; 48 tracts (29%) are classified as having medium to low social vulnerability; 20 tracts (12%) have medium to high social vulnerability; and 01 tract (1%) has high social vulnerability and 01 tract (1%) very high social vulnerability, both in the Alecrim neighborhood, as shown in graph 10:

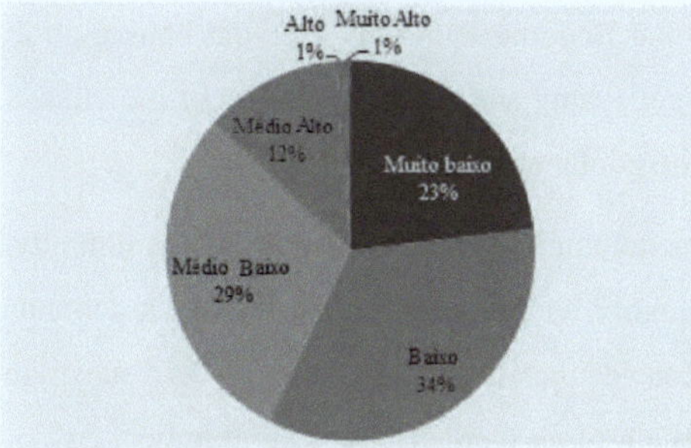

Graph 10: Distribution of census tracts according to degree of social vulnerability for the eastern zone. **Source:** Prepared by Marysol Medeiros based on survey data.

The Barro Duro community in Mae Luiza has incomplete paving in several streets, irregular construction on dunes - an area belonging to Environmental Protection Zone 10 (ZPA-10) - with steep slopes that favor the rapid flow of surface water. It is an area in the process of being occupied, adjacent to areas of consolidated occupation and with reasonable infrastructure.

The Alto da Colina community is also an area in the process of irregular occupation on the edges of ZPA-02 (Parque das Dunas), where many residents are abandoning their homes due to the high risk of landslides, given that some houses have already been interdicted by the Municipal Civil Defense.

Figure 14: Images of the Barro Duro and Alto da Colina/Mae Luiza communities. Left: unpaved street in the Barro Duro community and, in the background, occupation of ZPA-10. Right: Stairs giving access to the Alto da Colina community, which occupies the dunes on the edge of ZPA-02. **Source:** author's personal archive, May 2013.

Also in Mae Luiza, the Sopapo community has a high number of houses built on slopes and lowered floors. As a result of this situation, Rua Joao XXIII, the main road in the neighborhood, is on a lower level and the surrounding slopes are completely waterproofed, making it a point of constant flooding. Not only that, but houses on 2ª travessa Joao XXIII are on the verge of collapsing and the vulnerability of individuals is greater, as there are even houses with no adults present.

Between the Mae Luiza and Areia Preta neighbourhoods lies the Aparecida community, which is an area of profound social inequality with *non edificandi* sites, as it contains dunes that, for the most part, have not been adequately stabilized to serve as a substrate for the dense urban occupation that is taking place there. There is an imminent risk of the houses found between Atalaia Street and Guanabara Avenue collapsing, given the proximity of the buildings to the cut made in the land, making this area very unstable from a geomorphological point of view (figures 15 and 16).

Figure 15: Images of the Sopapo and Aparecida/Mae Luiza communities. Left: Houses occupying hillsides in the Aparecida community between Rua Atalaia and Avenida Guanabara, as well as the obvious social differentiation that exists there. Right: Stairs, or street, on one of the slopes in the Sopapo community. **Source:** author's personal archive, May 2013.

The following image shows the location of the points mentioned above in the Mae Luiza neighborhood, which is surrounded by dunes.

Figure 16: Location of the communities in the Mae Luiza neighborhood. On the left is the Barro Duro community; in the center, to the north, the Alto da Colina community; in the center, near the beach, the Sopapo community; on the right, the Aparecida community. **Source:** Google Earth, March 2014.

In the Alecrim neighbourhood, two factors played a role in two sectors being characterized as having very high SVI (sector 240810205080119)[6] and high SVI (sector 240810205080118). The first factor was the number of improvised dwellings which, according to the IBGE, are establishments that were not built for housing purposes and are exercising that function; the second factor refers to dwellings with no male residents and surroundings without drainage systems; however, in these sectors specifically, the variables that weighed most heavily were those linked to gender.

The Sao José do Jacó community, in the Rocas neighborhood, has been occupied for more than ten years. Basic urban services such as paving, drainage and sanitation are still precarious and found only occasionally in the community; it has many narrow alleys with no paving, an accumulation of garbage and cracked staircases that give access to the streets further up the hillside, as well as many precarious dwellings.

6 Census sector code number given by the IBGE.

Figure 17: Images of the Sâo José do Jacó/ Rocas community. Left: Street (alleyway) with no paving and surface drains. Right: staircase leading to the street above the hillside. **Source:** author's personal archive, December 2013.

The Passo da Pàtria community, located in the Cidade Alta neighbourhood, is in a densely occupied area with basic infrastructure. It is characterized by being a fishing community, but it is close to the Potengi River estuary and the Baldo Canal and is exposed to the risk of flooding. Despite the low income of the residents, the sector has undergone some intervention projects, such as the Passo da Pàtria Project of the Habitar Brasil Program, which has greatly improved the conditions around the sector.

Figure 18: Images of the Passo da Patria/ Cidade Alta community. Left: proximity of the houses to the Baldo canal. Right: boats of the fishermen who live in the Passo da Patria community. **Source:** author's personal archive, September 2011.

It should be noted that the sectors that include the communities of the Mae Luiza neighbourhood, the Sao José do Jacó community and the Passo da Pàtria community are classified by the IBGE as subnormal agglomerations, which are characterized by areas with the following features: "it is a group made up of at least 51 housing units lacking, for the most part, essential public services, occupying or having occupied, until a recent period, land owned by others and being arranged, in general, in a disorderly and/or dense manner." (IBGE, p.18, 2010). However, for this methodology, areas of social vulnerability do not necessarily coincide with subnormal agglomerations, given that much broader variables such as gender, income and schooling are taken into account when drawing up the SVI and, according to the IBGE:

> Subnormal settlements can fall into the following categories, depending on the criteria of urbanization standards and/or precariousness of essential public services: invasion, irregular or clandestine allotments, and invaded areas and irregular and clandestine allotments regularized in recent times. (IBGE, p.18, 2010).

The census tracts covering the communities of São José do Jacó, Sopapo, Alto da Colina, Aparecida, Barro Duro and Passo da Pàtria showed social vulnerability ranging from medium low to medium high. This is because the factor that most influenced this classification was the general characteristics of the residents, especially the variables linked to income, the number of illiterate people, gender and the density of residents per household. As these are areas of very old urban occupation, some sectors have certain basic urban services such as garbage collection, sewage disposal and paving. For this reason, the variables linked to the infrastructure around the homes did not carry as much weight compared to the sectors in the southern zone, where the lack of a drainage system predominated. What's more, the other factors did not show high social vulnerability, so that after combining the factors, the SVI in these subnormal agglomerations was not so high.

3.2.3 West Zone

The West Zone of Natal has ten neighborhoods (figure 19): Planalto, Guarapes, Cidade Nova, Felipe Camarao, Dix-Sept Rosado, Cidade da Esperança, Quintas, Bom Pastor, Nordeste and Nossa Senhora de Nazaré. This region has deep social problems, including a lack of income for many families, a lack of infrastructure and basic urban services, as well as high crime rates.

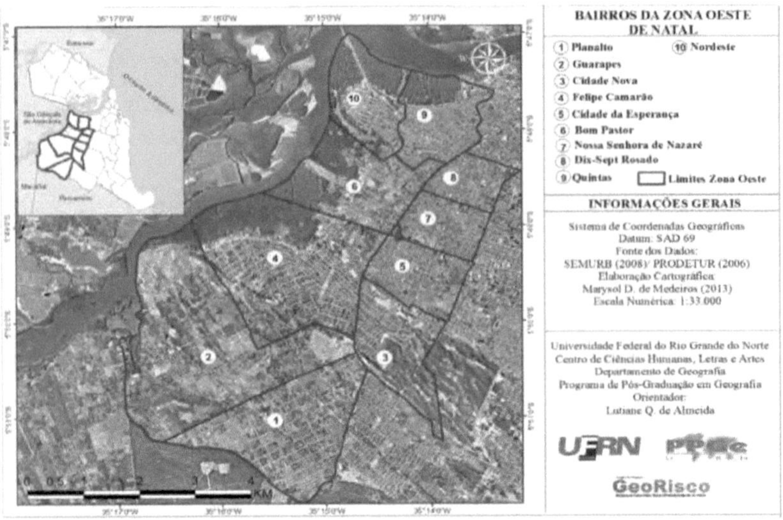

Map 19: Neighborhoods in the Western Administrative Region of Natal, RN. **Source:** Prepared by Marysol Medeiros using data from PRODETUR (2006) and SEMURB (2008).

The Planalto neighborhood has a large number of sectors with high or very high social vulnerability. It is a neighborhood in growing expansion and is developing in areas adjacent to areas of consolidated occupation with reasonable infrastructure, in addition to presenting increasingly heterogeneous economic patterns, where you can find houses in gated communities and many others built with scraps of wood and siding. It is located around the city's old garbage dump; most of its streets are unpaved, except for the main roads; it has no urban drainage works and it is very common to see garbage and debris piling up, especially in vacant lots. The images below illustrate this situation (figure 21).

Figure 20: Images of the Planalto neighborhood. Left: shacks and modest houses found in the neighborhood and in the background, condominiums showing the social differentiation in the area.

Right: marked with arrows, rubble and sewage flowing in the open causing laminar erosion in Rua Dom Antônio de Almeida Lustosa. **Source:** author's personal archive, May 2013.

In an area of very high social vulnerability, located in the Planalto neighborhood, on the border with the Guarapes neighborhood, is the Monte Celeste community. With houses made of mud or scraps of wood and siding, without paving or any other urban infrastructure, the community has existed for more than 11 years and is located around the city's old rubbish dump. Next to this sector is the community of Leningrado, in the Guarapes neighborhood, which has high levels of social vulnerability.

Figure 21: Images of the Monte Celeste/ Planalto community. Left: detail of the material used to build the houses. Right: houses in the community around a huge gully with an accumulation of garbage. **Source:** author's personal archive, September 2013.

The Cidade Nova neighborhood has a highly socially vulnerable sector known as Baixa do Câo. This sector is highly vulnerable because it lacks urban infrastructure, with unpaved streets and drainage, forming real sewage lagoons in the middle of the streets; poor social conditions for residents; precarious housing and lack of public lighting.

Figure 22: Images of the Baixa do Câo/ Cidade Nova community. Top left: modest houses built at the base of the dune. Top right: street with waterproofed soil and sewage (indicated by arrows) running under the open sky. Bottom left: Cabugi Street occupying dune slopes, without paving and showing erosion due to surface sewage runoff (indicated by arrows). Bottom right: Accumulation of sewage in Rua Cabugi, forming a lagoon in the middle of the street. **Source:** author's personal archive, September 2011.

The Guarapes neighbourhood is considered to be an area of peripheral expansion, far from the more urbanized core, with a low density of occupation and lacking basic infrastructure. Some streets are paved and have public lighting, but the majority of the streets do not have these facilities. One sector in particular is noteworthy: it is the largest census tract in the city[7] and, in turn, is highly socially vulnerable. Despite its large size, the sector is sparsely populated, with 701 households and around 2016 inhabitants according to data from the last demographic census. In this sector is the Anatália de Souza Settlement, which has existed for three years, with extremely precarious housing, made of mud or siding, as shown in the images below:

7 With the exception of the sector that includes Parque da Dunas, as there are no buildings there.

Figure 23: Images of the Anatalia de Souza/Guararapes settlement. Left: modest mud houses. Right: unpaved street and houses made from scraps of wood and canvas. **Source:** author's personal archive, May 2013.

Sectors with high social vulnerability can be found in the Felipe Camarâo neighborhood due to the economic deprivation of residents and urban facilities, as well as gender (with a predominance of women as heads of household, and houses with no male residents), in addition to the lack of a rainwater drainage system. The proximity of the homes in these sectors to the mangrove area makes the situation even more chaotic, since the sewage that runs off the surface of the higher streets is concentrated in the areas located on the lower topography and flows into the mangrove, which in many cases are in the backyards of several houses.

Figure 24 Images of the Felipe Camarao neighborhood. Left: street with sewage flowing into the Potengi river estuary. Right: accumulation of sewage from backyards flowing into the estuary. **Source:** author's personal archive, September 2011.

Between the Bom Pastor and Quintas neighborhoods is the Riacho das Lavadeiras or Riacho das Quintas, with sectors around it with medium to high social vulnerability, the most preponderant factors being the characteristics of the residents and the general characteristics of the surroundings, as these are areas that are home to a population with low incomes and living in areas without adequate urban infrastructure, with unpaved streets, an accumulation of garbage, sewage in the streets and no drainage equipment. In this same situation are the Parque 13 de Maio community in the Dix-Sept Rosado neighborhood, some sectors of the Felipe Camarao neighborhood and the Mosquito Community in the Nordeste neighborhood.

Figure 25: Images of the Parque 13 de Maio/Dix-Sept Rosado community. Left: unpaved street with open sewage drains, a source of contamination and erosion, as well as an accumulation of garbage. Right: laminar erosion caused by sewage runoff in an unpaved street. **Source:** author's personal archive, September 2011.

Figure 26: Images of the sectors around the Riacho das Lavadeiras/ between the Quintas and Bom Pastor neighborhoods. Left: one of the many alleyways found perpendicular to the Riacho das Lavadeiras, with sewage flowing in the open, a source of contamination and exacerbated soil sealing. Right: Accumulation of garbage in an unpaved road on the banks of the Riacho das Lavadeiras. **Source:** author's personal archive, September 2011.

Figura 27: Images of the Mosquito/Northeast community. Left: Paved street, with public lighting, but no rainwater drainage. Right: Accumulation of garbage in the backyards of houses on the estuary of the River Potengi. **Source:** author's personal archive, September 2011.

Of the 244 sectors in the west zone, 02 (1%) have very high social vulnerability and are located in the Guarapes and Cidade Nova neighborhoods. 19 sectors (8%) have high vulnerability and are located in the Planalto, Felipe Camarâo, Bom Pastor, Guarapes, Quintas and Cidade Nova neighborhoods. 36 (15%) have medium to high vulnerability, 91 (37%) medium to low, 71 (29%) low and 25 sectors (10%) very low. The next graph (graph 11) illustrates this breakdown of sectors according to degree of vulnerability.

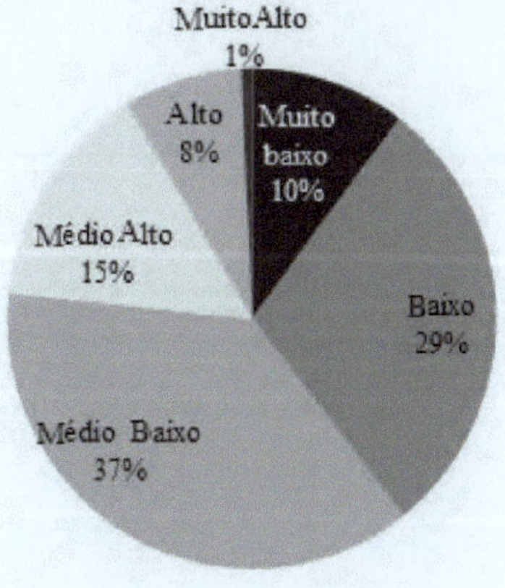

Graph 11: Distribution of census tracts according to degree of social vulnerability for the west zone. **Source:** Prepared by Marysol Medeiros based on survey data.

3.2.4 North Zone

Natal's North Zone has seven neighborhoods (figure 28): Salinas, Igapó, Potengi, Redinha, Pajuçara, Lagoa Azul and Nossa Senhora da Apresentaçâo. Like the West Zone, residents of the North Zone face constant problems due to the lack of public policies aimed at improving the quality of life for those who live in this region.

Of the 311 census tracts registered in the northern zone, only 05 (2%) have very low social vulnerability, 02 in the Redinha neighborhood, 02 in Lagoa Azul and one in Pajuçara. 71 sectors (23%) have low social vulnerability; 128 (41%) medium to low vulnerability; 82 sectors (26%) with medium to high vulnerability; 17 (5%) high and 08 (3%) with very high vulnerability, making it the administrative region of the city with the most socially vulnerable sectors.

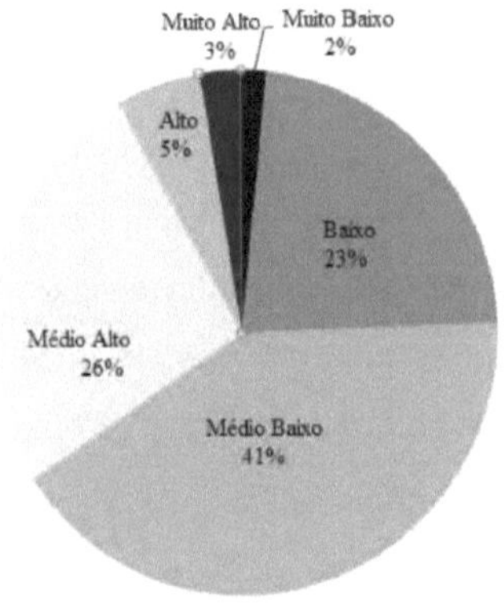

Graph 12: Distribution of census tracts according to degree of social vulnerability for the northern zone.
Source: Prepared by Marysol Medeiros based on survey data.

Figura 28: Neighborhoods in the Northern Administrative Region of Natal, RN. **Source:** Prepared by the author using data from PRODETUR (2006) and SEMURB (2008).

The neighborhood of Redinha has two sectors with very high social vulnerability and different vulnerability factors. The sector that includes the Redinha Public Market presented the characteristics of the residents as the main factors for this result, due to the high density of residents per home, homes without bathrooms and lack of adequate sewage disposal, as well as the low income of the families. Allied to these factors, the number of improvised homes was a significant factor. The adjoining sector showed high social vulnerability due to the "general characteristics of the residents" factor. The other sector with very high vulnerability is in the community of Africa, where the lack of urban infrastructure equipment makes the "characteristics of the surroundings of the homes" factor more relevant.

Presenting the same problems as the Africa community, the sectors encompassing the Salinas community have unpaved roads, a build-up of garbage and surface run-off of wastewater effluents; the factor linked to gender and the lack of rainwater drainage was also significant in configuring high vulnerability.

The sector with high social vulnerability located in the Pajuçara neighborhood has no infrastructure equipment, making the factors "characteristics of the surroundings of the homes" and "homes in areas without public lighting" more relevant. The factor that had the greatest impact on very high social vulnerability in the Lagoa Azul neighborhood was the lack of public lighting, followed by the precarious infrastructure of the homes and the poor urban conditions of the surroundings. The four sectors with very high social vulnerability in the Nossa Senhora da Apresentaçâo neighborhood have all the high vulnerability factors, except for improvised dwellings. These sectors include the Jardim Progresso community. The other sectors with high social vulnerability had factors such as "general characteristics of the residents" and "characteristics of the surroundings of the homes" with greater weight.

Based on the methodology used to draw up the Social Vulnerability Index, it was possible to see that the distribution of census tracts with the highest levels of social vulnerability was basically between the neighborhoods located in the western and northern areas of the city, with the economic conditions of the residents and the lack of infrastructure in both the homes and the streets being the main factors behind this result.

The sectors with high social vulnerability in the eastern and southern zones of Natal had factors linked to the lack of a rainwater drainage system and gender, given that they had households with no male presence and were located in areas with no rainwater drainage system.

Social vulnerability is an essential parameter in the diagnosis of risk areas, since from this analysis it is possible to characterize the social groups that are exposed to danger. However, it is essential to know which risks - in this case linked to natural hazards - the population is exposed to.

Therefore, the next chapter deals with the natural factors that make certain areas more vulnerable than others from an environmental point of view, in this case taking into account morphodynamic aspects.

4. NATAL'S ENVIRONMENTAL VULNERABILITY

This stage of the work consists of a diagnosis of the physical and natural factors that contribute to the occurrence of natural hazards in Natal, such as flooding and landslides. Therefore, a meticulous survey and analysis of primary and secondary data was carried out, in addition to the production of thematic maps that helped in this analysis.

4.1 Methodology

This work was based on the methodology proposed by Crepani (2001), developed from the concept of Ecodynamics developed by Tricart (1977), in which he argues that in areas where morphogenesis prevails over pedogenesis, erosive processes are more prominent and therefore modify the relief; on the other hand, pedogenesis favors the formation of soils. Thus, the dynamics of environments are greater where morphogenesis predominates, and they are much more unstable from the point of view of natural dynamics.

The methodology proposes an analysis of the physical aspects that make up the area by means of thematic maps that include pedological, geological, geomorphological, phytogeographic and land use information, which will then be superimposed on the rainfall intensity curves of the study area.

The maps drawn up containing the themes of the different physical aspects form the georeferenced Thematic Information Plans (PI). In this way, the vulnerability analysis was carried out on the basis of each PI separately, and then the vector intersection of these Information Plans was carried out, with the final result being the map with the Basic Territorial Units (PI UTB), which Crepani (2001) defines as being "a geographic entity that contains environmental attributes that allow it to be differentiated from its neighbors, while at the same time having dynamic links that articulate it to a complex network integrated by other territorial units". (CREPANI, 2001; ps. 12 and 13).

The author of the methodology proposes Vulnerability to Soil Loss values according to the susceptibility of each element to the development of morphogenesis. Therefore, it is only necessary to consult the predefined tables that have been proposed.

Vulnerability values ranging from 1.0 to 3.0 are assigned according to the susceptibility of morphodynamic processes. In total, it is possible to classify the study area on a scale with 21 vulnerability values. Areas with values close to 1.0 are dominated by pedogenesis processes; otherwise, if morphogenesis is the dominant process, the closer to 3.0 the area is, the more vulnerable it is to natural risks related to soil loss.

The following table (table 03) summarizes the 21 values in which the areas can be classified, and how they should be identified on the vulnerability map according to the color corresponding to the degree of vulnerability.

Table 04- Vulnerability scale for landscape units.

Landscape Units		Average		Degree of Vulnerability	Degree of saturation			
					Verm.	Green	Blue	Colors
U1		3,0		Vulnerable	255	0	0	
U2		2,9			255	51	0	
U3		2,8			255	102	0	
U4	V	2,7			255	153	0	
U5	U	2,6		Moderately vulnerable	255	204	0	
U6	L	2,5	E		255	255	0	
U7	N	2,4	S		204	255	0	
U8	E	2,3	T		153	255	0	
U9	R	2,2	A	Medium stable/ Vulnerable	102	255	0	
U10	A	2,1	B		51	255	0	
U11	B	2,0	I		0	255	0	
U12	I	1,9	L		0	255	51	
U13	L	1,8	I		0	255	102	
U14	I	1,7	D	Moderately stable	0	255	153	
U15	D	1,6	A		0	255	204	
U16	A	1,5	D		0	255	255	
U17	D	1,4	E		0	204	255	
U18	E	1,3		Stable	0	153	255	
U19		1,2			0	102	255	
U20		1,1			0	51	255	
U21		1,0			0	0	255	

Source: CREPANI, 2001

The following equation is used to assign vulnerability values to each territorial unit:

$$V = \frac{(G + R + S + Vg + C)}{5}$$

In which:

V: vulnerability

G: vulnerability for the geology theme

R: vulnerability for the geomorphology theme

S: vulnerability to the issue

pedology **Vg:** vulnerability

for the vegetation theme **C:**

vulnerability to the climate issue.

Crepani's methodology uses an arithmetic mean of five elements: geology, geomorphology, pedology, vegetation and climate. However, the Land Use PI is not taken into account in the calculation and is only superimposed on the map resulting from the information obtained by the above equation. However, for this work, it was necessary to adapt it, in view of the need to incorporate the forms of land use into the Vegetation PI, since the spatial area in focus has areas of strong anthropogenic influence.

4.2 Environmental characterization of Natal through the Information Plans

This subchapter will show the physical-natural characteristics of Natal, since in order to recognize a natural hazard it is necessary to distinguish which factors contribute to the development of the danger. The aim was to characterize the geomorphological and geological features, soil types and land use, as well as the climatic characteristics of the area studied.

4.2.1 Weather

The climatic characteristics of a region will directly influence the weathering of rocks and consequently the formation of soils; not only that, the climate controls the types of vegetation that will develop in a given region, but it also has a direct influence on the modification of the relief.

The climate is a factor responsible for triggering various natural hazards, whether they are immediate, such as the passage of a storm, or long-term, such as droughts and dry spells. However, most of the natural events that cause damage in Brazil are linked to the intensity of rainfall, in which the high amount of rainfall in a short space of time causes flooding, waterlogging and landslides due to the saturation of the soil.

For the reasons mentioned above, a characterization of the climate of the city of Natal will be carried out and consequently an analysis of the vulnerability of this factor will be carried out, taking into account the criteria chosen and listed in the methodology.

According to Koppen's classification, Natal has a hot humid climate, and Vianello and Alves (1991) classify the climate of the Potiguar capital as tropical rainy hot with a dry summer. The highest rainfall is recorded during the winter, concentrated in the months from April to June, according to the Rain Atlas of Brazil prepared by the Mineral Resources Research Company (CPRM).

The following table, organized from data provided by the Agricultural Research Company of Rio Grande do Norte (EMPARN), illustrates the distribution of rainfall in the city (table 06).

Table 05. Average Rainfall in Natal between 1984 and 2011

Year/ Months	Jan	Feb	Sea	Apr	May	Jun	Jul	Aug	Set	Out	Nov	Ten	Total
1984	48,9	88,3	247,4	393,4	623,7	247,1	140,1	190,7	39,6	48,9	14,4	19	2101,5
1985	144,3	256,1	352,5	177,3	354	191,6	394,8	181,3	127,6	1,4	19,9	23,3	2224,1
1986	52,5	238	284,3	577,3	355,9	316,9	112,5	194	81	57,3	127,7	41,1	2438,5
1987	26,4	76,9	246,8	280,9	87	349,3	281,7	36,7	50,8	20,7	15,5	5,4	1478,1
1988	18,7	127,4	332,4	277,9	370	404	309	128,8	78,3	9,7	21	81,5	2158,7
1989	38,9	35,1	70,3	341,2	137,3	88,8	102,9	121,9	50,5	39,5	24,4	104,5	1155,3
1990	42,5	23,6	58,9	141,4	187,9	275,7	167,2	195,1	61	23,1	54,5	2,7	1233,6
1991	27,1	16,4	106,1	306,5	304,9	234,6	143,5	73,4	9,4	24,4	45,5	6,8	1298,6
1992	52,6	184,4	288,7	315	130,8	343,4	80,5	157,7	32,1	4	26,2	6,4	1621,8
1993	21,6	29,4	84,8	227,8	39,9	164,9	129,6	35,8	10,9	61,9	15,5	36,9	859
1994	94,5	86,2	237	275,7	279,7	624,9	399,6	84,6	57,3	4,5	20,4	19,9	2184,3
1995	24,6	60,6	278	237,4	392,3	254,4	408,3	34,8	21,6	4,2	41,3	0,4	1757,9
1996	37	69,7	216,2	359,5	154,6	251,4	145,8	125,4	86,7	78,4	45,1	17,2	1587
1997	13,3	72,8	159,6	256,2	340	77,7	79,8	121,3	5,1	3	2,7	55,8	1187,3
1998	49,2	78,3	81,4	74,7	161,7	210	789	138,7	19	13,7	8,5	17	1641,2
1999	12,6	143,8	139,2	175,8	289,5	131,5	31,1	49,8	51,1	14,5	5,4	66,9	1111,2
2000	40	79,9	114,9	177,4	230	577,2	482,3	288,8	205,1	8,7	12,1	23	2239,4
2001	30,1	6,8	133,8	360,2	14,4	373,4	145,2	103,2	28,4	13,7	17,1	50,2	1276,5
2002	108,9	82,3	483,1	137,7	122,9	405,6	225,2	312,9	1	29,4	98,9	18,8	2026,7
2003	84,1	184	312	133,4	230,8	244,3	183,5	49,6	41,6	21,7	16,1	22,5	1523,6
2004	383,9	283	252	167,8	160,7	642,9	393,4	90,1	44,4	13	10,1	4	2445,3
2005	2	36,4	186,3	144	548,2	761,3	126,9	134,4	43,9	31,6	1,2	10,4	2026,6
2006	4,2	87,2	157,4	427,9	115,3	375,1	133,3	90,3	45,2	13,4	83,5	49,8	1582,6
2007	86,3	67,6	260,9	236,8	115,7	560,4	191,4	95,8	46,2	19,6	45,2	16	1741,9
2008	68,2	22,4	279,8	417	198,3	542	473,1	401,1	37,8	32	9,5	0,4	2481,6
2009	162,1	245,7	220,6	364,3	372,3	304,5	338,5	229,2	76,7	0,9	6,7	9,5	2331
2010	71,3	81,4	69,9	191	262,7	153,6	150,5	94,3	40,5	8,2	13,2	54,7	1191,3
2011	326,1	76,5	146,7	368,3	413	445,4	169,5	118	33,3	20,8	44	21,9	2183,5
Average	74,0	101,4	207,2	269,4	249,8	341,1	240,3	138,5	50,9	22,25	30,2	28,1	1753,2

Source: Prepared by Marysol Medeiros from EMPARN data.

Rainfall in the city of Natal is concentrated between April and July, with June being the wettest month, with an average rainfall of 341.14mm. This is due to the action of the Intertropical Convergence Zone, which is responsible for the rainfall in the state from January to April, and the East Waves, especially in June and July. On the other hand, the driest months are concentrated in spring, between September and November, with

October being the driest month in the Potiguar capital, registering only 22.22 mm of rain on average. The average annual rainfall in the city is 1753.15 mm.

The city's temperature range is very low, with mild temperatures, with an average annual temperature of 26.0°C, an average maximum temperature of 29.3°C and an average minimum temperature of 22.5°C.

The intensity and direction of the winds greatly influence the occurrence of landslides in dune areas, since the friable material (sand) is easily transported by the action of the wind, thus promoting the deflation of the topmost layer.

The intensity of the winds in the city reaches 5.27 m/s on average and the months of September and October record the highest averages, however, when it enters the first month of the rainy season, the intensity of the winds decreases to 3.17m/s in the month of April, as shown in the following graph. For most of the year, the winds act in a south-easterly direction (south-easterly trade winds - coming from the Atlantic Ocean).

As a coastal city with a high amount of available surface water and its proximity to the equator, Natal has high evaporation rates and consequently high relative humidity, as shown in the following graph, with monthly averages of no less than 74%. The relative humidity ranges from 74.4% in October (the driest month) to 81.3% in June (during the city's rainy season).

The driest periods of the year in the city correspond to the months with the most sunshine. In this case, the months with the most sunshine in Natal are between October and January, with 300.3 hours/month in November.

The methodology applied for this work suggests using only rainfall data, from which the rainfall intensity of the area is calculated by dividing the average annual rainfall in millimeters by the length of the rainy season in months (CREPANI, 2001, p 95). According to the methodology, the value of rainfall intensity is equivalent to the potential energy available in the system to be transformed into kinetic energy capable of causing erosion.

In this way, the rainfall intensities of the city of Natal were mapped using the rainfall isopleths available in the Rain Atlas of Brazil prepared by the CPRM. The isopleth values were divided by the number of months with the highest rainfall concentration in the city, which runs from April to July, giving a total of four months. The following maps show the amount of average rainfall in different parts of the city, as well as the degree of vulnerability of each area according to its rainfall intensity.

The following map shows that the capital city of Rio Grande do Norte has the highest rainfall totals near the coast and that these totals decrease as you move towards the mainland.

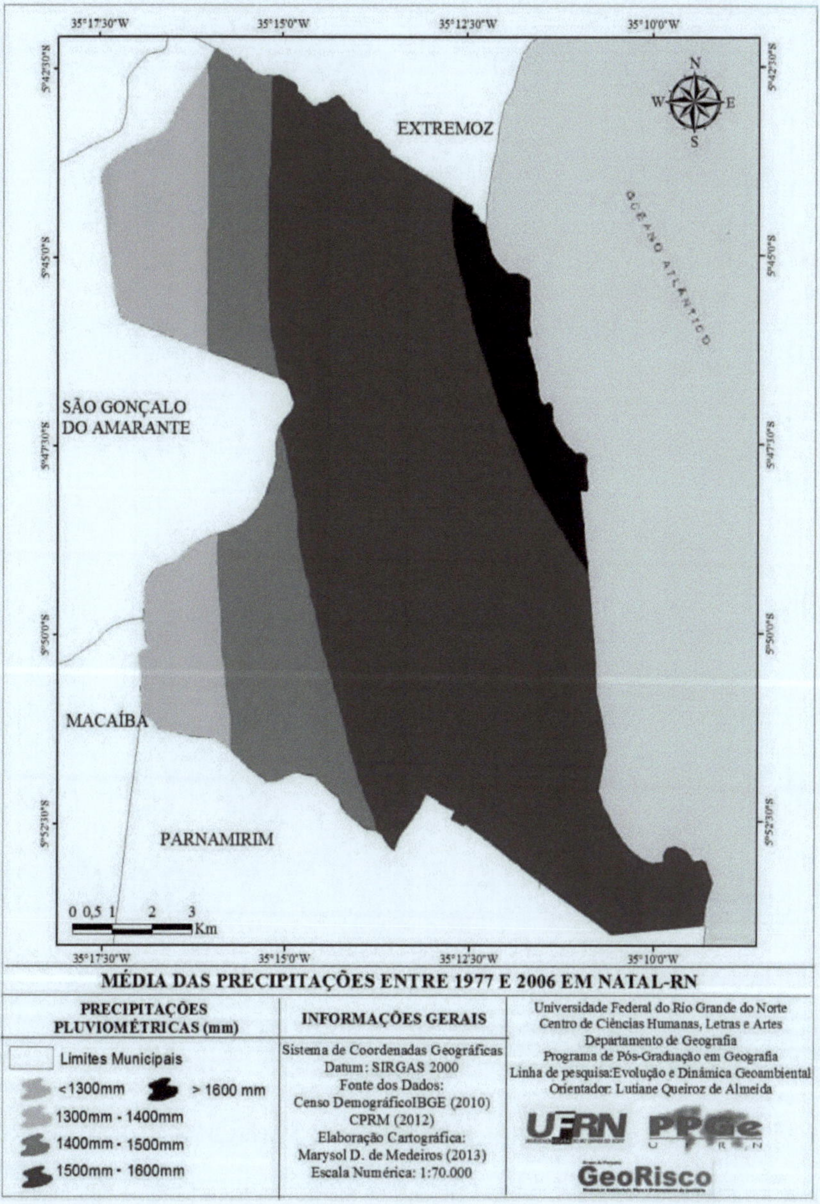

Figure 30: PI Climate. **Source:** Prepared by Marysol Medeiros from CPRM data (2012).

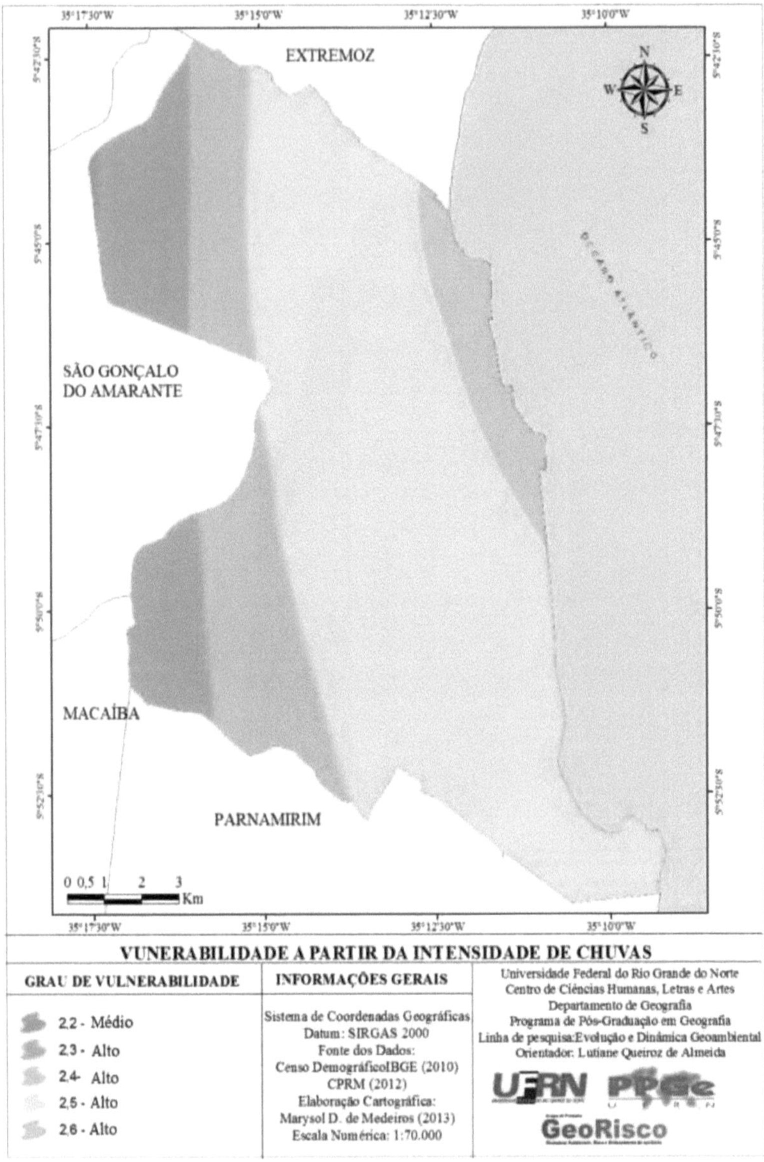

Figure 31: Climate Vulnerability. **Source:** Prepared by Marysol Medeiros based on data from the CPRM (2012).

Practically all the areas are highly vulnerable in relation to rainfall intensity. Only a small area to the northwest of the city has a Medium degree of vulnerability.

4.2.2 Geology

For this methodology, the most relevant geological information is related to the lithological characteristics that make up the substrate of the area in question. In this case, the degree of rock cohesion, porosity and fractures, as these are crucial pieces of information in determining susceptibility to morphogenesis in the face of weathering. This is because poorly cohesive rocks are more susceptible to erosive processes.

In order to draw up the maps containing the geological characterization and, consequently, the environmental vulnerability in terms of geology, data from CPRM's digital maps at a scale of 1:100,000 were used, which were then processed using the ArcGIS 9.3 geographic information systems program.

The capital of Rio Grande do Sul lies on sedimentary cover from the Cenozoic Era, mainly from the Neogene Period. The following recent deposits are found in this period:

- **Coastal Beach Deposits (N4lp):** these are formed by quartz sands with a grain size ranging from fine to very coarse, well sorted, rich in bioclasts and with the presence, in some places, of heavy minerals. These deposits are arranged in a narrow strip parallel to the coastline and originate from the processes of interaction between the tides and the coastal plain.

- **Non-vegetated coastal aeolian deposits (N4eln):** these are made up of fine to medium, whitish, well-sorted quartz sands with rounded grains. They can take a variety of forms, but are mainly frontal or parabolic, running through vegetated dunes. Their main characteristic, as the name suggests, is the absence of vegetation, which makes it difficult for the sediments to settle,

making them susceptible to wind transport.

- **Vegetated Coastal Aeolian Deposits (N4elv):** are composed of well-selected quartz sands; mature; reddish in color, due to the presence of iron oxides, to brownish due to the presence of decomposing organic matter; in the form of residual dunes or sandy sheets. It originates from the wind transport of sediments, which are then fixed by the vegetation that acts as a bulkhead. It is basically found throughout the city, covering most of the South, East and West administrative regions and a considerable portion of the north-east of the North region.

- **Channel Alluvial Deposits (N4ac):** made up of medium to coarse quartz sands; possibly conglomeratic; with cross-stratification and discontinuous levels of finer sediments (silty-clayey), especially towards the mouths (CPRM, 2012). In Natal, this unit is found along the valley of the Pitimbu River, in the southwestern part of the city and originates from subaqueous traction processes, encompassing channel facies and fluvial channel bars.

- **Fluvial-marine deposits (N34fm):** influenced by rivers and tides; made up of fine sands, silts and finely laminated clays; rich in carbonates and living, bio-detrital organic matter. They include mangrove deposits and tidal flats. Their presence in the capital of Rio Grande do Norte is along the Potengi River fluvial-marine plain.

- **Sandy and** sandy-clay **deposits (N4ar):** these are unconsolidated sandy to sandy-clay coverings of medium to coarse grain size. Locally, they may contain quartz fragments and/or remobilized sandstone pebbles (CPRM, 2012). These deposits are significantly located in the northwestern portion of the city's northern administrative region and in a small sample in the southwestern part of the western administrative region, between the Guarapes and Planalto neighborhoods.

- **Marine and Continental Deposits (N3mc):** according to CPRM mapping (2012), these are formed by medium to fine sandstones; with colors ranging from golden yellow to brown and orange; arranged discordantly over the Barreiras Group or ancient alluvial deposits. The discordance with the Barreiras Group is evidenced by the reworked ferruginous concretions of an Iateritic paleosol. It is located in a small area in the extreme northwest of the city, in the northern zone, more precisely in the neighborhood of Nossa Senhora da Apresentaçâo.

- **Old Alluvial Deposits (N3a):** made up of conglomeratic sandstones and conglomerates with quartz pebbles and fragments of various rocks, including ferruginous sandstones from the Barreiras Group. Subordinately, there are sandy clay sediments to medium sandstones of various colors. They show weak diagenesis and frequent paleoseismic structures. They are associated with the migration of the Potengi/Jundiai River paleochannel (Angelim, et al., 2007).

- **Sandy reefs (N4ra):** will not be included in this study due to their location (on the foreshore). The reefs are formed by carbonate cemented sandstones with bioclasts, medium to coarse grain size, distributed as reef lines parallel to the current coastline and are widely found along the city's coastline.

- **Barreiras Group (ENb):** had a sedimentation interval between the Paleogene and Neogene periods. It is made up of conglomeratic to conglomeratic sandstones, sandstones and claystones. It may have oxidation horizons, giving it a color that varies between bank, red and purple. It is possible to find sediments from the Barreiras Group arranged along the coast, forming trays and cliffs.

The following figure (figure 32) shows the distribution of lithological units in the city of Natal.

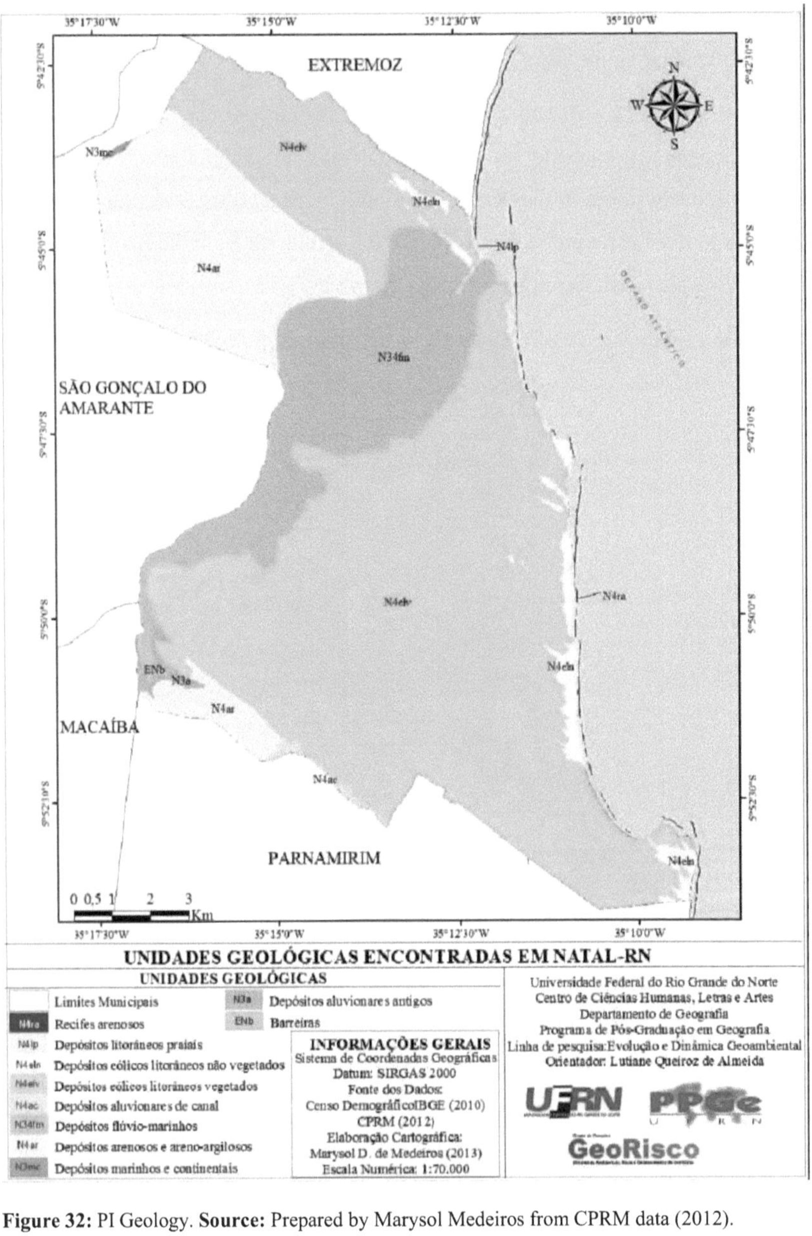

Figure 32: PI Geology. **Source:** Prepared by Marysol Medeiros from CPRM data (2012).

According to the methodology, each type of rock has a vulnerability value according to its degree of resistance. Based on the weights given to each type of rock, the vulnerability map for the geology theme was drawn up. The table below shows the

Vulnerability to Denudation scale for the most common rocks.

Table 06: Vulnerability scale to denudation of the most common rocks.

Rocks	Average	Rocks	Average	Rocks	Average
Quartzites or Metaquartzites	1,0	Milonites, Muscovite quartz, Biotite, Chlorite schist.	1,7	Quartz sandstones or orthoquartzites	2,4
Rhyolite, Granite, Dacite	1,1	Pyroxenite, Amphibolite, Kimberlite, Dunite	1,8	Conglomerates, Subgrauvacas	2,5
Granodiorite, Quartz Diorite, Granulites	1,2	Hornblende, Tremolite, Actinolite shale	1,9	Grauvacas, Arcozios	2,6
Magmatites, Gneisses	1,3	Schist staurolite, Granatiferous schists	2,0	Siltstones, Argillites	2,7
Phonolite, Nepheline Sienite, Trachyte, Sienite	1,4	Phyllite, Matassiltite	2,1	Brochures	2,8
Andesite, Diorite, Basalt	1,5	Slate, Matargillite	2,2	Limestones, Dolomites, Marls, Evaporites	2,9
Anorthosite, Gabbro, Peridotite	1,6	Marble	2,3	Unconsolidated sediments: Alluvium, colluvium, etc.	3,0

Source: CREPANI (2001), adapted by MEDEIROS (2014).

Figure 33 shows the distribution of the degrees of Vulnerability of the rocks of the lithological units in the city of Natal and according to their cohesive capacity.

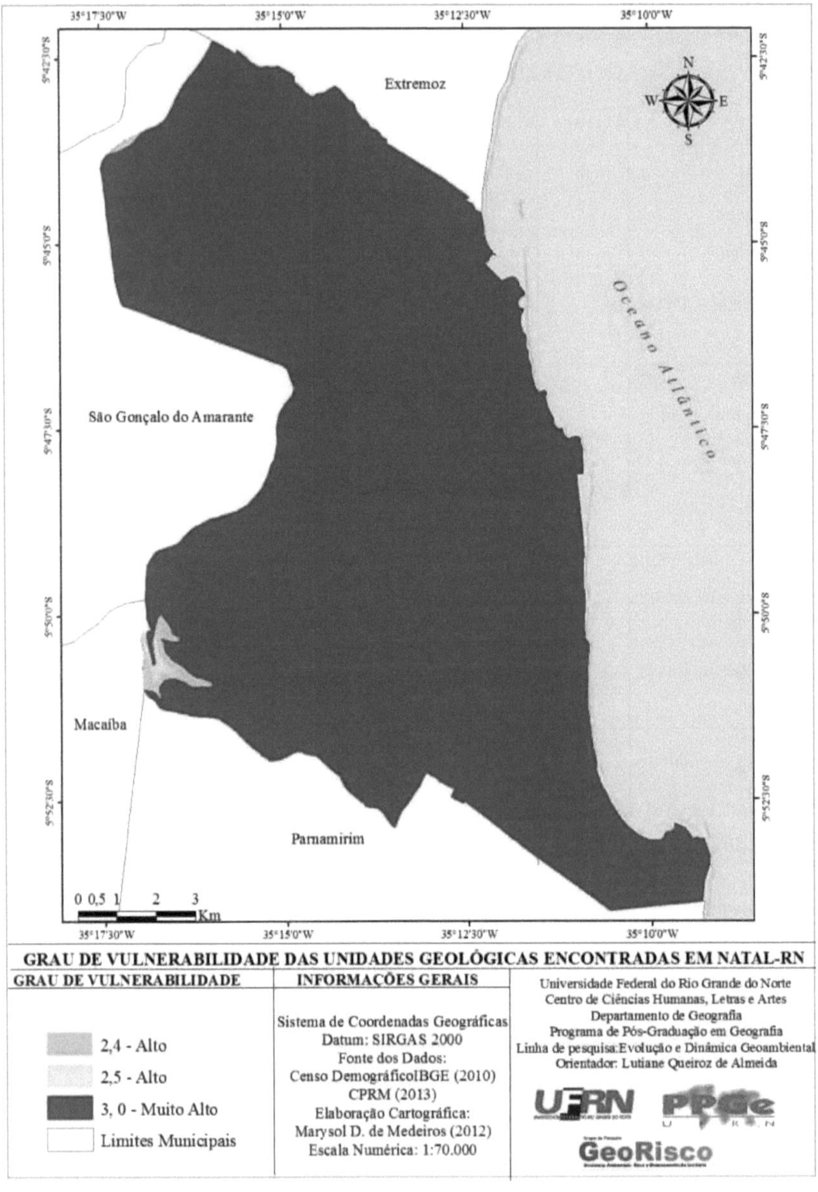

Figure 33: Geological vulnerability. **Source:** Prepared by Marysol Medeiros based on data from the CPRM (2012).

The substrate of the city of Natal is made up of sedimentary rocks, most of which are unconsolidated. The Coastal Beach Deposits (N4lp), Non-Vegetated Coastal Eolian

Deposits (N4eln) and Vegetated Coastal Eolian Deposits (N4elv) are composed of very friable material that is easily transported by wind action, making these geological units with a vulnerability value equal to 3.0, therefore classified as very high vulnerability.

The Channel Alluvial Deposits (N4ac), Fluvial-Marine Deposits (N34fm) and Sandy and Sandy-Argylic Deposits (N4ar), which have a finer grain size with the presence of clay, are also classified as vulnerable and with a vulnerability score of 3.0, as they are unconsolidated sediments that are easily eroded.

The Marine and Continental Deposits (N3mc) and the Ancient Alluvial Deposits (N3a) are basically made up of sandstones, which are sedimentary rocks generally made up of cemented sands, and therefore show greater cohesion between the particles that make up their lithological material. As such, these two units are considered to be highly vulnerable, with a value equivalent to 2.4.

The Barreiras Group (ENb) is made up of conglomerate material, with the presence of rounded rock fragments connected by a cement which, in the case of the capital of Rio Grande do Sul, may be iron oxide or hardened clay. The Barreiras Group is considered to be a unit with a high degree of vulnerability due to the cohesiveness of its particles, however, the cementation it presents makes it more resistant and consequently with a degree of vulnerability equal to 2.5.

For some sedimentary deposits, especially sandy ones, the action of water alone is not the main exogenous agent modifying the relief. In the case of the area under study, the poor consolidation of the lithological material means that many deposits are transported by aeolian processes. However, they present serious structural limitations to urban occupation, with the possibility of house collapses and landslides during more concentrated rainfall.

The table below summarizes the lithological components that make up the geological units found in the Potiguar capital.

Table 07: Summary of rocks and vulnerability values

Geological Units	Lithological composition	Weight
Coastal Beach Deposits (N4lp)	Quartz sands	3,0
Non-Vegetated Coastal Wind Deposits (N4eln)	Quartz sands	3,0
Vegetated Coastal Wind Deposits (N4elv)	Quartz sands	3,0

Channel Alluvial Deposits (N4ac)	Quartz sands	3,0
Fluvial-marine deposits (N34fm)	Fine sands, silts and clays	3,0
Sandy and sandy-alluvial deposits (N4ar)	Unconsolidated sandy to sandy-clay deposits	3,0
Marine and continental deposits (N3mc)	Sandstones	2,4
Old Alluvial Deposits (N3a)	Conglomeratic and ferruginous sandstones	2,4
Barrier Group (ENb)	Conglomerates	2,5

Source: Prepared by Marysol Medeiros

4.2.3 Geomorphology

As far as geomorphology is concerned, the methodology uses morphometry, taking into account the altimetric range and slope more than the analysis of morphography. This is because the altimetry and slope of the terrain determine whether there is enough potential energy - potential energy is directly linked to the force exerted by gravity - for material to flow down a slope. Therefore, the higher the altitude, the greater the potential energy for descent and, the steeper the terrain, the less stability the material will have, so that high altitudes combined with steep slopes will make it easier and faster for materials to descend the slopes, making this environment more unstable.

The altimetry and slope maps were drawn up using a topographic survey (contour lines) with an equidistance of one meter between the isolines, provided by Natal City Hall and acquired using aerial images obtained from the Northeast Tourism Development Program (PRODETUR). By processing the data using the Arcgis 9.3 geographic information system, it was possible to determine the morphometry of the city.

Following the methodology, the altimetry can be divided into 21 Vulnerability classes according to altitude, i.e. the higher the slope, the greater the degree of vulnerability. However, the city of Natal has a maximum altitude of 120 meters, so 12 vulnerability classes were established, as can be seen in the table below.

Table 08: Vulnerability values for altimetry

Altimetric Amplitude (m)	Vulnerability / stability	Altimetric Amplitude (m)	Vulnerability / stability
<20	1,0	67,5 - 77	1,6
20 - 29,5	1,1	77 - 84,5	1,7

29,5 - 39	1,2	84,5 - 94	1,8
39 - 48,5	1,3	94 - 103,5	1,9
48,5 - 58	1,4	103,5 - 113	2,0
58 - 67,5	1,5	113 - 122,5	2,1

Source: CREPANI (2001), adapted by MEDEIROS (2014).

The greater the altimetric range, the greater the potential energy involved in the system, causing the friable or unconsolidated material found in the higher parts to acquire more kinetic energy during its descent and, consequently, greater erosive capacity, due to the speed with which it travels down the slope from the higher areas to the lower ones. Conversely, the lower the altimetric range, the speed at which the material descends will be less erosive and, consequently, pedogenesis will be favored in this environment.

Natal's relief is basically composed of: fixed and mobile dunes, which can be found superimposed on the coastal trays; and the coastal trays, in turn, are mainly seen as cliffs. The lowest-lying areas of the city are located along the Potengi River's river-marine plain, which divides the northern region of the city from the other administrative regions; the coastal plain along the coast; the Pitimbu River's river plain in the south and the Doce River in the north; and the lake plains, the most significant of which are the lagoons located in the north of the city.

As shown in figure 34, the highest altitudes in the city are found along the fixed dunes, which are deposits of marine and/or continental origin remolded by winds and fixed by vegetation. The fixed dunes with the highest altitudes are mainly located in Parque das Dunas (Environmental Protection Zone/ ZPA-02); Parque da Cidade (ZPA- 01); near the old landfill site in Cidade Nova (ZPA-04) and in the dunes that form Morro do Careca (ZPA-06).

The coastal tablelands are characterized by relatively low altitudes, ranging from 30m to 100m; they are associated with sedimentary soils, in the case of Natal with the Barreiras Group. Its relief is flattened and, in the capital of Rio Grande do Norte, it is covered by sandy deposits and fixed and mobile dunes. Their edges are steep and form living or inactive cliffs. The alluvial plains of the Pitimbu River and the Potengi River estuary are located on the coastal plateau.

The Potengi River is influenced by the sea for 20km from its mouth, forming a river-marine plain. This plateau has low altimetry levels of up to 10 meters, making it a constantly floodable area, favoring the development of the mangroves that cover the entire length of the Potengi River's fluvial-marine plateau.

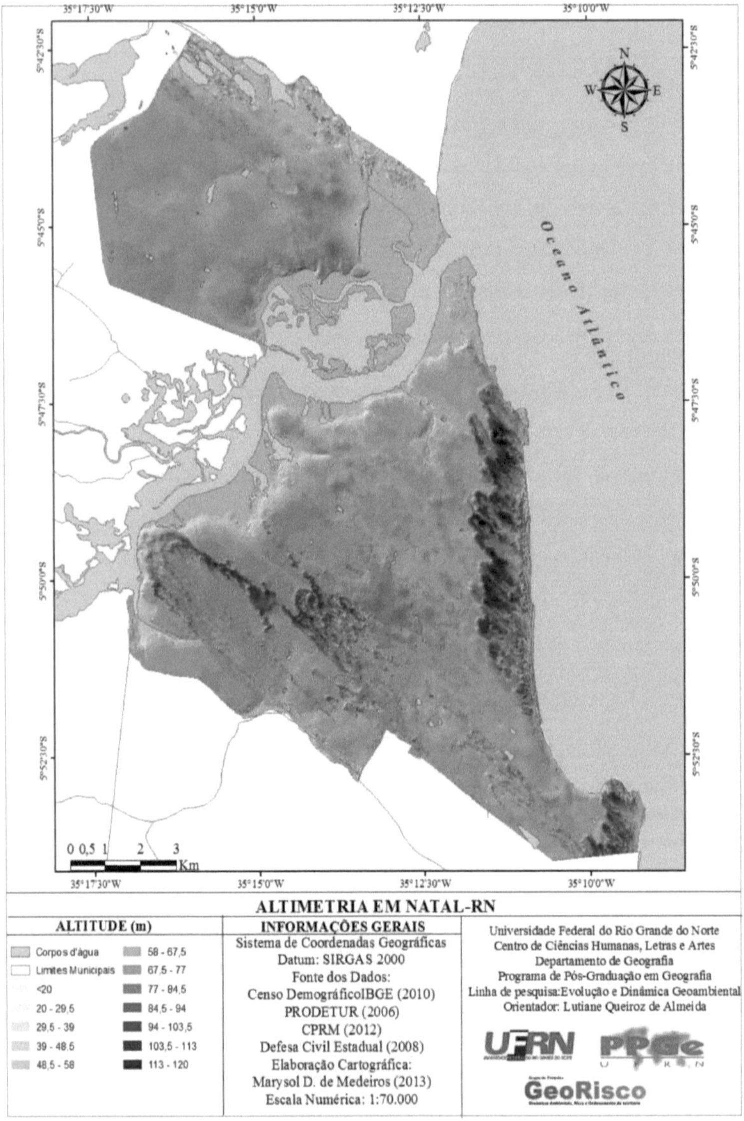

Figure 34: Natal's hypsometry. **Source:** Prepared by Marysol Medeiros based on data from CPRM (2012), PRODETUR (2006).

Flowing into the Potengi River are the Riacho das Quintas, also known as the Riacho das Lavadeiras, which is small and less than half a kilometer long; and the 4.7 km Riacho (or canal) do Baldo[2] in the eastern part of the city. This canal crosses important avenues in the capital, including Avenida Prudente de Morais and Avenida Rio Branco, which are densely occupied and have a high level of waterproofing. Due to recurrent flooding, the stream has been channelized into a trapezoidal shape, but there are still episodes of flooding, especially downstream in the Passo da Pàtria community.

The Pitimbu River is smaller than the Potengi, but it contributes to the water supply in the south, east and west of the city. It lies to the south of Natal, between the Planalto and Pitimbu districts, bordering Natal and the municipality of Parnamirim. It belongs to the hydrographic basin of the Pirangi River and is fed by groundwater rising from the sediments of the Barreiras Group (MEDEIROS, 2001).

The Doce River is also not long, at around 14km, nor is it voluminous, but the heavy occupation of its banks has de-characterized the riparian forest and contributed to the silting up of its bed. This river is fed by surface water from the Extremoz lagoon and, like the Pitimbu river, by groundwater from the Barreiras. Since the slope is lower than the static level of the groundwater, during the rainy season the volume of water increases considerably, causing flooding.

Trapped between the interdunal ridges are the city's natural lagoons, which are outcrops of the Dunas/Barreiras aquifer or which arise from high rainfall. The heavy urban occupation of the areas surrounding the lagoons seals the soil, increasing the amount of water flowing from the surface and, consequently, causing the lagoons to overflow, contributing to flooding.

According to Medeiros (2001), the city has areas that form closed depressions, making it possible to form confined sub-basins, where rainfall that runs off the surface is directed there, forming flooded areas. An example of these depressions is Avenida Salgado Filho, near Praça de Mirassol, which for several years, during rainy periods, accumulated large amounts of water, becoming a lagoon. Only with urban drainage works has this inconvenience been remedied.

When drawing up the vulnerability map for altimetry (figure 35), high altitudes were taken into account, as well as flood plains as being of very high vulnerability, since plateaus, areas subject to flooding and with intense morphodynamics, are highly unstable. In order to delimit the areas of flooding, we adopted the criteria used by Medeiros (2001), who states that the lagoons in the capital of Rio de Janeiro are located at altitudes between 5.0 and 29.0 meters and Pereira (2012), who found that areas with altitudes below 29.0 meters in the capital of Rio de Janeiro have a recurrence of flooding when rainfall is over 200mm/month.

Slope acts as a preponderant factor and is defined as the inclination of the relief in relation to the horizon (GUERRA, 2008). In order to understand how slope acts as a preponderant factor in morphogenesis, it is first necessary to understand that every body is influenced by various forces. For there to be sliding, for example, the sum of the forces acting on matter must be non-zero, since for movement to occur, the resistance force must be smaller. Therefore, the greater the angle of inclination, the greater the force of resistance against gravity and other opposing forces must be in order to keep the system in balance.

Crepani (2001) explains that the steeper the slope, the faster the transformation of potential energy into kinetic energy of rainwater, and thus the greater the speed of the water masses and their capacity for transportation, which is why morphogenesis prevails.

As can be seen in figure 36, the city of Natal has low levels of slope over most of its territory, at no more than 6%. However, in the portions where the interdunar cordons are found, the slope can exceed 50%, especially on the leeward slopes, which are highly unstable.

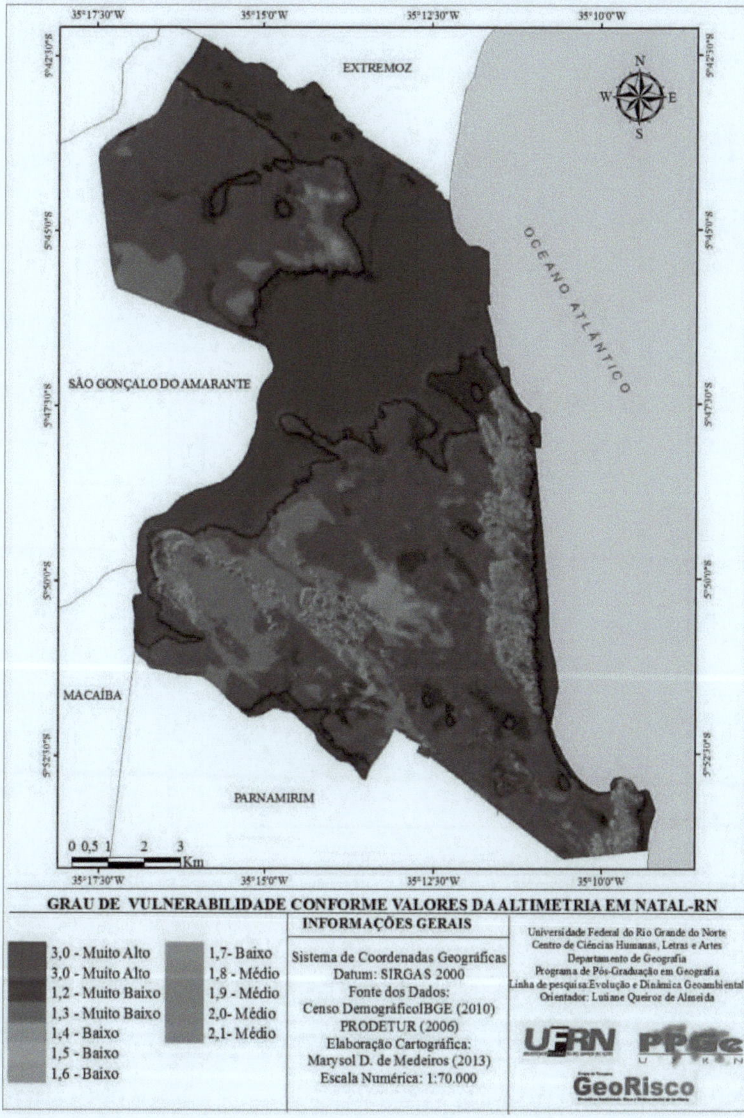

Figure 35: Altimetry vulnerability**. Source:** Prepared by Marysol Medeiros based on data from CPRM (2012), PRODETUR (2006).

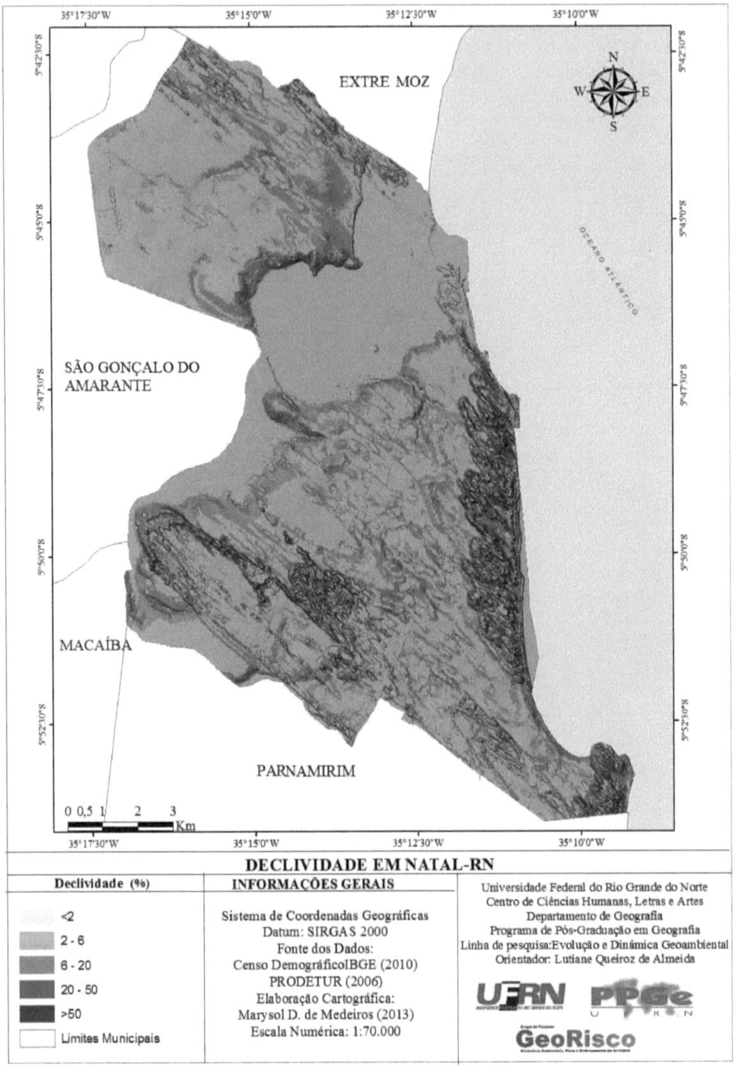

Figure 36: Natal's slope. **Source:** Prepared by Marysol Medeiros based on data from CPRM (2012), PRODETUR (2006).

To create the Natal slope vulnerability map (figure 37), the criteria in the following table were adopted, based on the assumption that areas with small slope angles - less than 2% - would have a very low vulnerability value, i.e. equivalent to 1.0, with pedogenesis prevailing. On the other hand, areas with slope angles greater than 50% were classified as having very high vulnerability, with a value of 3.0.

Table 09: Slope classes with their respective vulnerability scale values

Morphometric classes	Slope (%)	Vulnerability values
Very low	<2	1,0
Bass	2 - 6	1,5
Medium	6 - 20	2,0
High	20 - 50	2,5
Very high	>50	3,0

Source: CREPANI (2001), adapted by MEDEIROS (2014).

In order to draw up the geomorphology vulnerability map, an arithmetic mean was made of the slope and altimetric amplitude variables, based on the overlapping of the maps that had been drawn up previously.

Using the arithmetic mean of the slope and altimetry values, the geomorphological vulnerability map was created (figure 38). From this it can be inferred that, despite the relatively low altitude of the city's geomorphological features, the high slope values found, especially in the dunes, were responsible for the high vulnerability of these areas. On the other hand, the portions with altimetric levels that are prone to rainwater accumulation or flooding, which were classified as very high vulnerability in terms of altitude, had their degree of vulnerability lowered to low vulnerability, since, based on the overlapping of the maps, it is possible to see that they are plateaus and therefore have a low degree of slope.

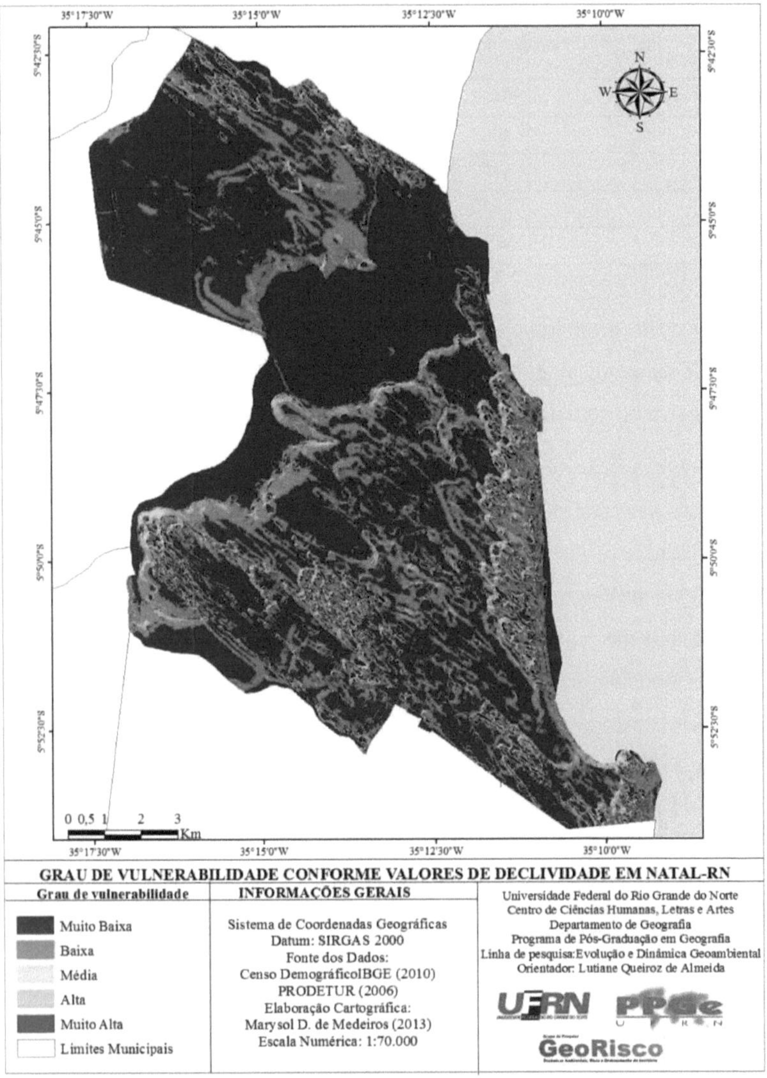

Figure 37: Slope Vulnerability. **Source:** Prepared by Marysol Medeiros based on data from CPRM (2012), PRODETUR (2006).

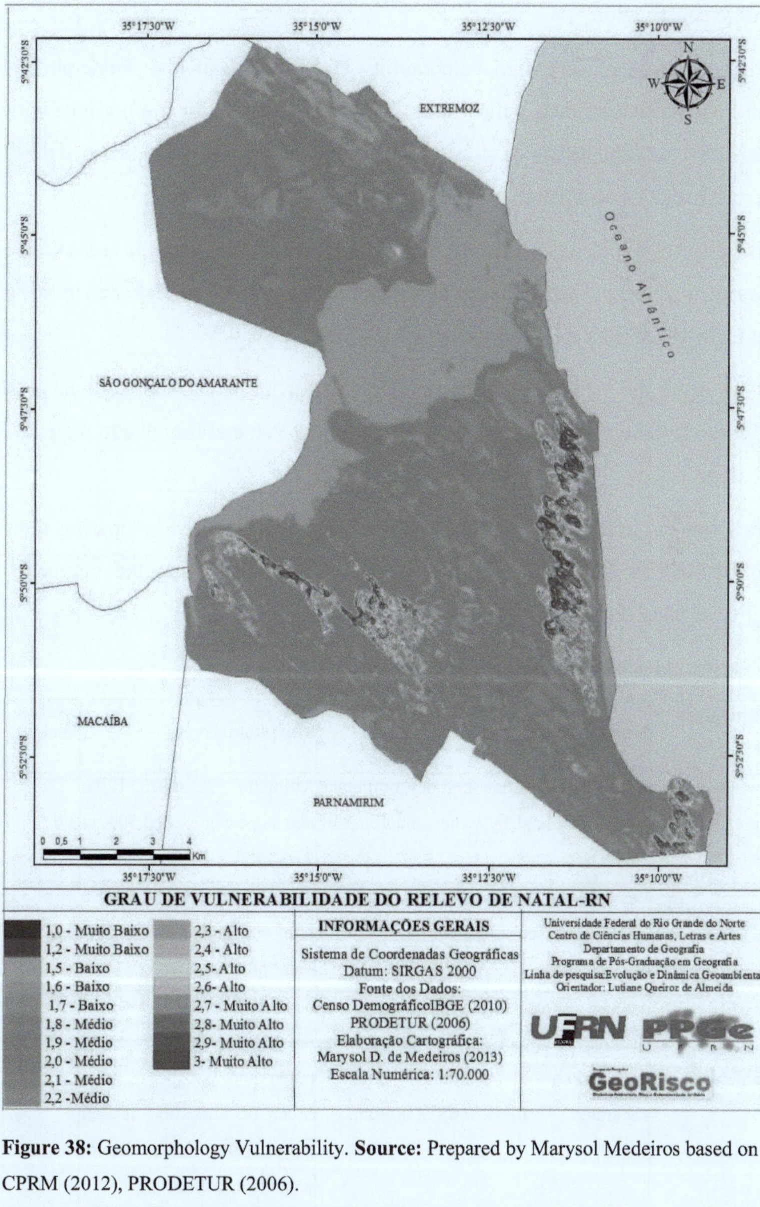

Figure 38: Geomorphology Vulnerability. **Source:** Prepared by Marysol Medeiros based on data from CPRM (2012), PRODETUR (2006).

Where the city's vegetated dunes are located are the areas with the highest degrees of vulnerability, which can reach a value equivalent to 3.0. However, these are areas with low or no occupation, as they are the city's environmental protection zones, but they are not free from the urban pressure exerted on the surroundings, with some invasions and inappropriate construction.

Surrounding the areas of very high vulnerability are the areas of high vulnerability, which can be identified by their yellow and lime green color. These are portions with lower altimetry levels and less steep slopes compared to the former, but in many of these dunes there is irregular occupation.

Most of the city has a medium degree of vulnerability, coinciding with the areas where there is heavy urban densification, as the dunes that once existed were flattened to make way for the growth of the capital of northern Rio Grande do Sul.

On the other hand, the lacustrine, fluvial-marine and fluvial plains, despite their high vulnerability to flooding, were classified as being of low vulnerability due to their low degree of slope.

The table below summarizes the main landforms found in the city, comparing their morphografic and morphometric characteristics, as well as scaling the degree of vulnerability of these units.

Table 10: Summary of relief units and vulnerability values

Geomorphological Units	**Morphography**	**Morphometry**	**Weight**
Fixed Dunes (Adf)	Deposits of marine and/or continental origin remolded by winds and fixed by vegetation.	Altimetry can reach 120m and slope can exceed 50% to leeward.	3,0
Dunes (Ad)	Deposits of marine and/or continental origin remolded by winds.	Altimetry can reach between 85m and 90m. Slope >20% and <50%.	2,9
Coastal Tablelands (t51/t52)	Tabular forms: flat-topped relief, with different orders of magnitude and depth and depth of drainage, generally separated by valleys and flat bottoms.	Deepening drainage equal to 51 (very poor) and 52 (poor), with order of dissection greater than 3,750m and less than or equal to 12,750m.	1,5
Marine Fluvial Plain (Apfm)	Flat areas resulting from the combination of fluvial and marine accumulation processes, generally subject to periodic flooding with mangrove vegetation.	Altimetry not exceeding 10m and slope >6%.	2,5

River Plateau (Apf)	Flat area resulting from fluvial accumulation subject to periodic flooding. Occurs in valleys with alluvial fill.	Altimetry not exceeding 45m and slope >6%.	2,5
Lake Plain (Apl)	Flat area resulting from lake accumulation processes, comprising lakes, sandy ridges and marginal dykes.	Altimetry not exceeding 45m and slope >6%.	2,0

Source: Prepared by Marysol Medeiros based on data from the RADAMBRASIL Project (1981).

4.2.4 Pedology

Based on morphodynamics, the methodology takes into account the maturity of the soil, since older soils are more decomposed and consequently deeper, allowing for greater stability against erosive agents. Other relevant factors are soil structure, clay content and permeability.

Pedology will be a preponderant factor in classifying an area as Vulnerable in this methodology, since soil characteristics will indicate the degree of morphodynamics of the landscape units, showing whether morphogenesis prevails over pedogenesis or the opposite.

In order to map the types of soils in the city of Natal at the scale desired in this work, information obtained from IDEMA-RN was used, which has data on a scale of 1:500,000. This data was processed in the GIS program Arcgis 9.3, and it was necessary to adapt this original mapping using vectorization with the help of image interpretation from the *Worldview-2* satellite (2010), altimetric data with isolines of one meter of equidistance imposed by PRODETUR and data from the wells drilled by the Water and Sewage Company of Rio Grande do Norte (CAERN) throughout the city.

The development of soils requires the action of the climate and living beings on the source material (which can be mineral or organic); the relief as a conditioner of stability so that the material can develop *in situ,* and time, knowing that it takes many years for a mature type of soil to form. These factors mean that each type of soil has intrinsic characteristics that will have a direct impact on other aspects of the natural landscape, such as the type of vegetation, water infiltration capacity and, consequently, erosion processes.

As can be seen in figure 39, the northern capital of Rio Grande do Sul has the following types of soil:

Quartzarenic neosols (RQ), which are characterized by not maintaining lithic contact up to a depth of 50 cm; they have an A-C horizon sequence; with coarse or fine sands, essentially quartz, up to a depth of at least 150 cm. Despite their young age, these soils are extremely deep, non-hydromorphic and excessively drained. These soils are associated with flat, gently undulating, undulating and strongly undulating terrain and are found forming fixed or mobile dunes, as well as the beach strip.

Fluvial neosols (RY) are soils formed from the deposition of unconsolidated river sediments of varying granulometry. They are poorly developed, but can vary in depth from moderately deep to very deep. Fluvial neosols have a diverse morphology and can vary locally, mainly due to the soil's material of origin. They are soils with imperfect or moderate drainage and indiscriminate texture, and can therefore present a glei horizon, pale-colored horizons or horizons with characteristic reduction mottling. These soils are found on river banks and are linked to flat terrain.

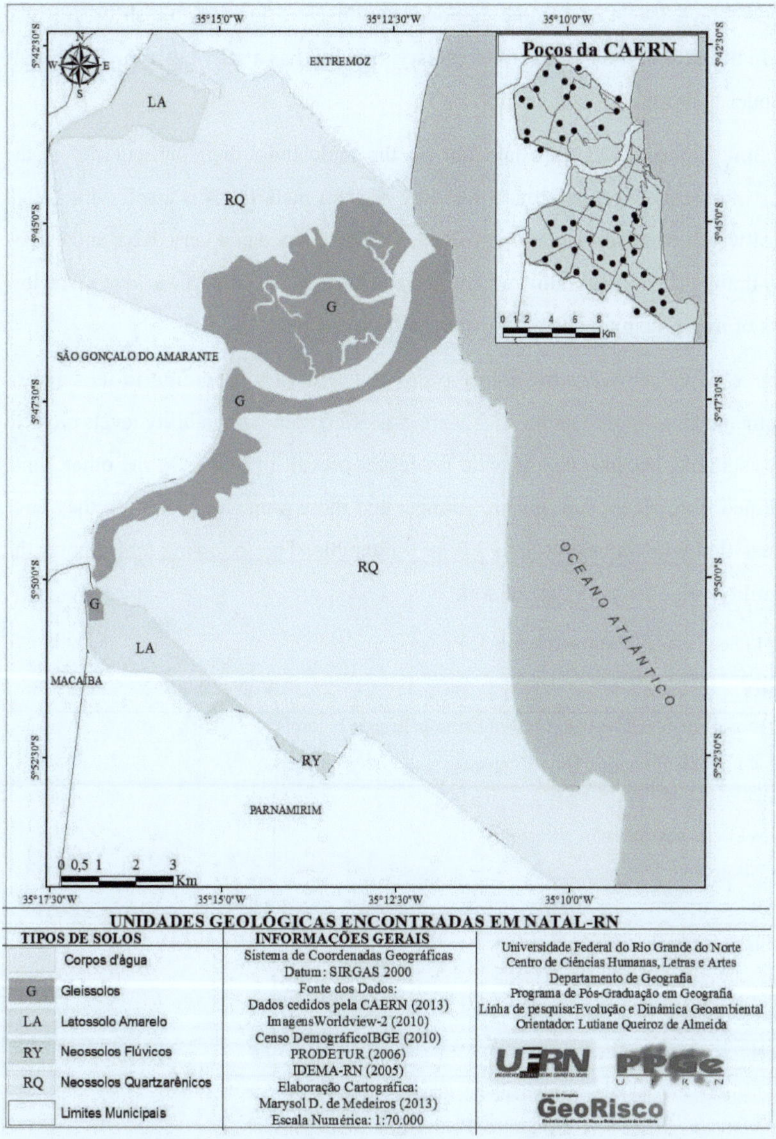

Figure 39: PI Pedology. **Source:** Prepared by Marysol Medeiros based on data from CAERN (2013), PRODETUR (2006), IDEMA-RN (2005), *Worldview-2* images (2010).

The Gleissolos (G) found along the Potengi river estuary are characterized by being hydromorphic soils with permanent water accumulation (as they are poorly or very poorly drained), leading to strong gleaning due to iron reduction processes and the presence of high levels of organic carbon, resulting in a greyish, bluish or greenish colour. In the case of the Potengi river estuary, this soil has a salty horizon. These soils occur under hydrophilous vegetation.

The Yellow Latosols (AL) are laid out on the tablelands; their particularity is the presence of a brassolic B horizon, formed by mineral material. It is a soil with a high degree of weathering, consequently evolved, well developed and very deep, and can be strongly drained. They are yellow with B horizon hues of 7.5YR or more, due to the low amounts of iron (goethite) and aluminum (gibbsite) oxides.

The degree of vulnerability for each type of soil will vary according to its stage of development. Thus, more mature soils were classified with vulnerability levels close to 1.0, i.e. as stable, because pedogenetic processes prevail in them. On the other hand, soils formed from recent deposits are younger and more prone to erosion, so they have been classified with values close to 3.0, or vulnerable. The following table shows the vulnerability value for each type of soil.

TABLE 11 - Soil vulnerability/stability values

Soil classes	**Vulnerability**
Yellow Latosols Red-Yellow Latosols Red Latosols Brunos Latosols (...) Humic Latosols Bruno (...) Humic Latosols	**1,0**
Argisols Argissolos Luvissolos Alissolos Nitossolos Argisols Nitosols Luvissolos Chernosols Planosols Spodosols	**2,0**
Cambisols	**2,5**
Litholic neosols Fluvial neosols Regolithic neosols Quartzarenic neosols Vertisols Organosols Gleissolos Plintossolos Gleissolos Plintossolos Rocky outcrops	**3,0**

Source: Crepani (2001) adapted by Marysol Medeiros

Based on the values in the table above, it was possible to draw up the pedology vulnerability map as shown in figure 40.

With the soil types shown in figure 39, it can be seen that the study area contains soils with classifications equivalent to the two extremes of vulnerability according to the methodology.

The most stable soils and those classified as being of very low vulnerability (with a vulnerability value of 1.0) in the city of Natal are the Yellow Latosols, because they are porous soils, generally with a sandy texture and permeability that guarantees the percolation of water, making them strongly drained and leached soils. They are mature and their yellow color is due to the absence of iron oxides.

Gleissols, Quartzarenic Neosols and Fluvial Neosols are considered to have very high vulnerability, being unstable soils with a value of 3.0. They are poorly developed soils with very different characteristics.

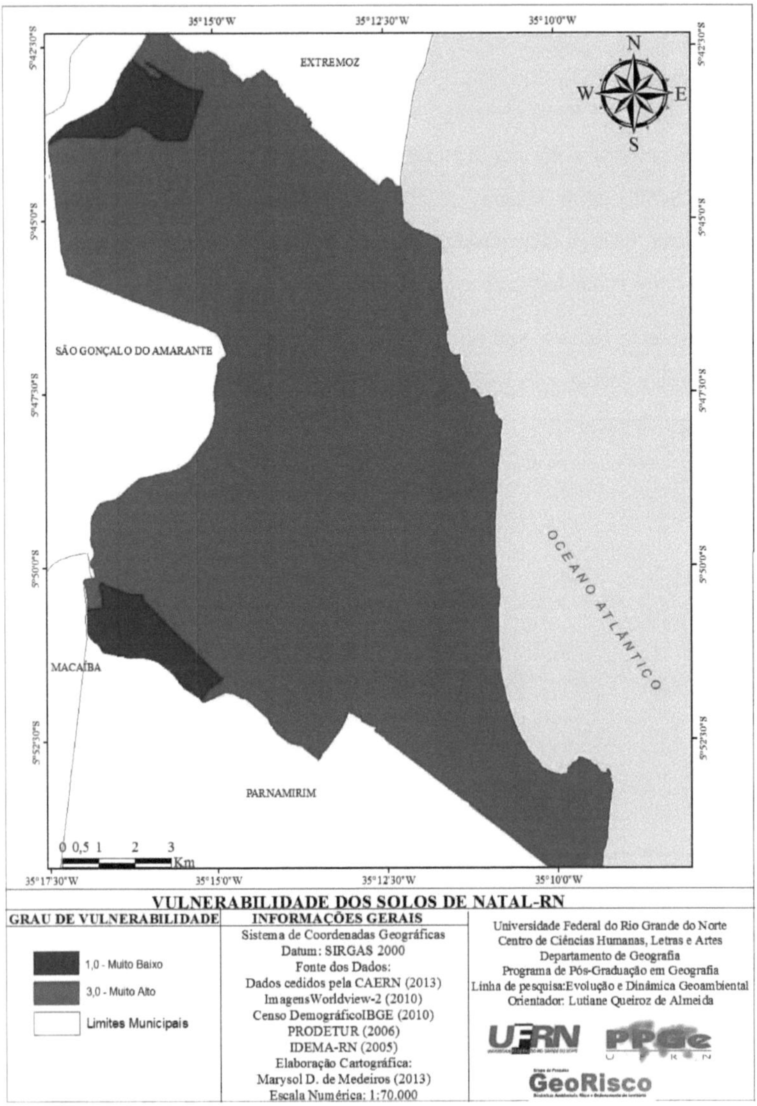

Figure 40: Pedological vulnerability. **Source:** Prepared by Marysol Medeiros based on data from CAERN (2013), PRODETUR (2006), IDEMA-RN (2005), *Worldview-2* images (2010).

Quartzarenic neosols are excessively drained due to their sandy texture, which allows for high porosity, as they are essentially composed of quartz grains with little aggregation between the particles, as well as a small amount, or even absence, of organic matter and clay. They are deep soils, but the friable material that makes them up makes them susceptible to erosion, especially through wind transport.

Fluvial neosols are poorly developed, hydromorphic soils formed by the superimposition of recent alluvial sediments in layers with no pedogenetic relationships between them. Because they are soils that can be easily transported by the action of water, they have been classified as very vulnerable. The Gleissolos, like the Fluvial Neosols, are poorly drained, poorly developed and spend most of the year in contact with water, being easily transported by hydrodynamic processes, therefore classified as very vulnerable.

The following table summarizes the degree of vulnerability and the main characteristics for each type of soil mapped in the area in question.

Table 12: Summary of soil types and respective vulnerability values

Types of soil	**Main features**	**Weight**
Yellow Latosol (LA)	Mature, deep, leached, porous and permeable	1,0
Quartzarenic neosol (RQ)	Young, disaggregated particles, absence of organic matter and clay, excessively porous and easily eroded.	3,0
Fluvial neosol (RY)	Poorly developed, poorly drained, shallow.	3,0
Gleissols (G)	Poorly developed, excessively poorly drained, shallow.	3,0

Source: Prepared by Marysol Medeiros.

4.2.5 Vegetation and land use

Vegetation cover acts as a soil fixative, since plant roots are able to prevent the movement of soil lumps. In addition to this factor, exposed soils are more vulnerable to the actions of weathering agents, since tree canopies act as a protective layer, preventing direct solar radiation from reaching the soil, encouraging the development of organic matter and avoiding the direct dripping of rainfall that can cause the *"splash"* effect that breaks up soil particles. Vegetation prevents soil compaction, promotes soil aeration, permeability and porosity and hinders concentrated surface runoff, inhibiting erosion.

The Vegetation and Land Use map was made using image processing in the GIS Arcgis 9.3 program, in which polygons representing the different uses and types of land cover were vectorized with the help of image interpretation from the *Worldview-2* satellite (2010) - which has a spatial resolution of up to 0.5 meters. The main vegetation occurrences that could be analyzed on a mapping scale equivalent to 1:50,000 were selected, as well as the main forms of land use. The mapping carried out by the

Municipal Civil Defense of the city's risk areas was superimposed on this vectorization.

As can be seen in figure 41, the city of Natal has several lagoons, either natural ones, which rise from the groundwater table, or artificial ones for collecting rainwater from urban drainage.

The old Cidade Nova garbage dump is located in the western region, occupying the dunes between the neighborhoods of Cidade Nova and Felipe Camarâo. This area has been designated as such since 1968, when the Natal City Council already disposed of the city's solid waste in this open-air site. It wasn't until 1983 that the Cidade Nova landfill was set up, in which there were only the characteristics of a controlled landfill, but the lack of material to cover the waste led to the destruction of the surrounding dunes. According to data from Natal City Hall, solid waste was disposed of on a site of just 30 hectares decades ago, with the landfill of approximately 4 million tons of garbage by 1997, which formed a layer of garbage between 10 and 18 meters high. Since 2004, the area of the former Cidade Nova dump has served as a transshipment point, where the collected waste is temporarily dumped until its final destination at the controlled sanitary landfill in Cearâ-Mirim. However, the large volume of waste (around 700 tons/day) means that the situation at the old dump has not changed, so that today it is still possible to find scavengers, a lack of material to cover the waste and a lack of protective fences. This area has no capacity to receive more waste, since practically 100% of the site's capacity has already been used and the layer of old garbage is thick, preventing the construction of a new cell. In addition, it is located in an area of growing urban density and is arranged over dunes, where the leachate from the waste is easily absorbed by the soil, thus contaminating the aquifer (PREFEITURA DO NATAL, 1997). The following images illustrate the above.

Figure 41: Images of the dune at the Cidade Nova/ Cidade Nova - Planalto dumpsite. On the left,

indicated by the arrow, the transshipment area of the Natal landfill, the former Cidade Nova dump, and next to the dune, the construction of a residential condominium. On the right: the dune that houses the old Cidade Nova dump with a considerable layer of rubble, and at the foot of the dune, a concentration of garbage that is carried by the wind (due to the lack of protective screens) or by scavengers. **Source:** author's personal archive, May 2013.

In the 1970s, the Shrimp Project was created with the aim of promoting shrimp farming in the state. As a result, this economic activity began to occupy the mangrove areas along the estuaries in Rio Grande do Norte. It was no different with the Potengi River estuary, where the mangroves, especially in the northernmost part, were devastated to make way for shrimp larvae breeding ponds.

Densely urbanized areas, as the name implies, are those with the presence of urban facilities, with high population and residential density, and are usually accompanied by areas with intense soil sealing, either due to paving or excessive construction. Areas in the process of occupation are so called because they are areas adjacent to portions of consolidated occupation with reasonable infrastructure or areas of peripheral expansion, far from the urbanized core, with low density of occupation and lacking basic infrastructure. The risk areas listed here are those mapped by the Municipal Civil Defense and which generally have houses with precarious engineering conditions, as well as areas of irregular occupation or high urban density and little infrastructure.

With regard to the floristic types found in the city, the following can be mentioned:

Dense Ombrophilous Forest (Atlantic Rainforest) which is found in the city only in Parque das Dunas. Its characteristic is that it grows in regions with temperatures above 25°C and high rainfall (as can be seen on the map of the distribution of rainfall in the city of Natal, this vegetation is found in an area with average rainfall of over 1600mm per year), and it can also be known as rain forest. In Natal, the Ombrophilous Dense Lowland Forest is found, as it is a forest formation located at sea level, on the coastal plain, with altitude variation between 5 and 100 meters, covering tablelands of the Barreiras Group. The main floristic genera found are *Ficus, Alchornea, Tabebuia* and *Tapirira guianensis* (IBGE, 1991).

Figure 42: Remnants of the Atlantic Forest in the city of Natal. On the left: a residence occupying the edge of the City Park, with the Ombrophilous Dense Lowland Forest in the background. On the right: houses occupying an area where the Atlantic Forest once stood. **Source:** Personal archive, September 2013.

The Pioneer Formations are those Aorist types that develop on pedologically unstable substrates due to the constant sedimentation of the terrain. This group includes the Sclerophyllous Forest (Arboreal or Shrubby Restinga), which is influenced by the sea and is responsible for fixing the dunes along the coast, beaches and occasionally the Barreiras Group soils, which are covered by Quartzarenic Neosols. Its size varies from arboreal to herbaceous. According to the IBGE (1991), the restinga is made up of trees and a few smaller trees, with well-developed and irregular tops, and the species *Anacardium occidentale* (cashew tree) is easily found.

What basically changes from an arboreal restinga to a shrub restinga is the size of the vegetation; in them you can find the dominance of caméAtas arranged either spaced out or more densely over the discontinuous herbaceous-graminoid stratum. In general, the restinga is made up of species with sclerotized leaves, long and well branched roots. Herbaceous sandbanks, on the other hand, are found close to the sea and vary in density, with the main representatives being *Ipomoea pes-caprae* (beach parsley) and *Paspalum vaginatum* (beach grass) (IBGE, 1991).

The physiognomy of this vegetation is quite heterogeneous, with various strata (as can be seen in image 26). In the higher parts of the dunes, it is possible to find woody, shrubby types interspersed with herbaceous species; on the slopes of the dunes, there are often shrubby species arranged densely, to the point where the tops meet; at the foot of the dunes, the shrubby individuals are sheltered. The main species found in the tree layer are: *Apuleia Ieiocarpa* (Jitai), *Anacardium occidentale* (Cashew), *Hancornia speciosa* (Mangabeira), *Coccoloba latifólia* (Cashew), *Ceasalpina echinata* (Brazilwood), *Spondias tuberosos* (Umbuzeiro).

Figure 43: Schematic profile of the physiognomies of the areas of marine influence (Restinga): 1) arboreal; 2) shrub; 3) herbaceous. **Source:** RADAMBRASIL (1981).

Mangroves, whether arboreal or herbaceous, are also part of the Pioneer Formations. This vegetation is exclusive to brackish environments with a fluvial-marine influence, with individuals adapted to high levels of salinity, and the species Laguncularia *racemosa (*White mangrove), *Rhizophora mangle* (Red mangrove) and *Avicennia Germinans* (Canoe *mangrove*) are easily found along the banks of the Potengi river estuary. The vegetation is evergreen, with an average height of between 6 and 10 meters.

Figure 44: Mangrove swamp in the Potengi river estuary. On the left: mangrove near the old bridge, in the community of Mosquito, Nordeste neighborhood. On the right: mangrove in the Felipe Camarâo neighborhood. **Source:** Personal archive, September 2011.

The Varzea Fields are found along rivers and lakes, where fresh water floods the surrounding soils, leaving them wet all year round, with a diverse physiognomy, sometimes sparse, sometimes dense, and heights between 0.3 meters and 2 meters. The carnauba-free riparian forest can be found along the Pitimbu and Doce rivers and occupies a strip no longer than a few tens of meters; in some stretches it is dense and lush, evergreen, and its canopy can reach between 8 and 15 meters and some trees can reach 25 meters.

Sandy surfaces are basically related to dunes without vegetation and, with a few exceptions, exposed soils that can be found throughout the city. Figure 45 shows the different types of vegetation and uses found on the substrate in the city of Natal.

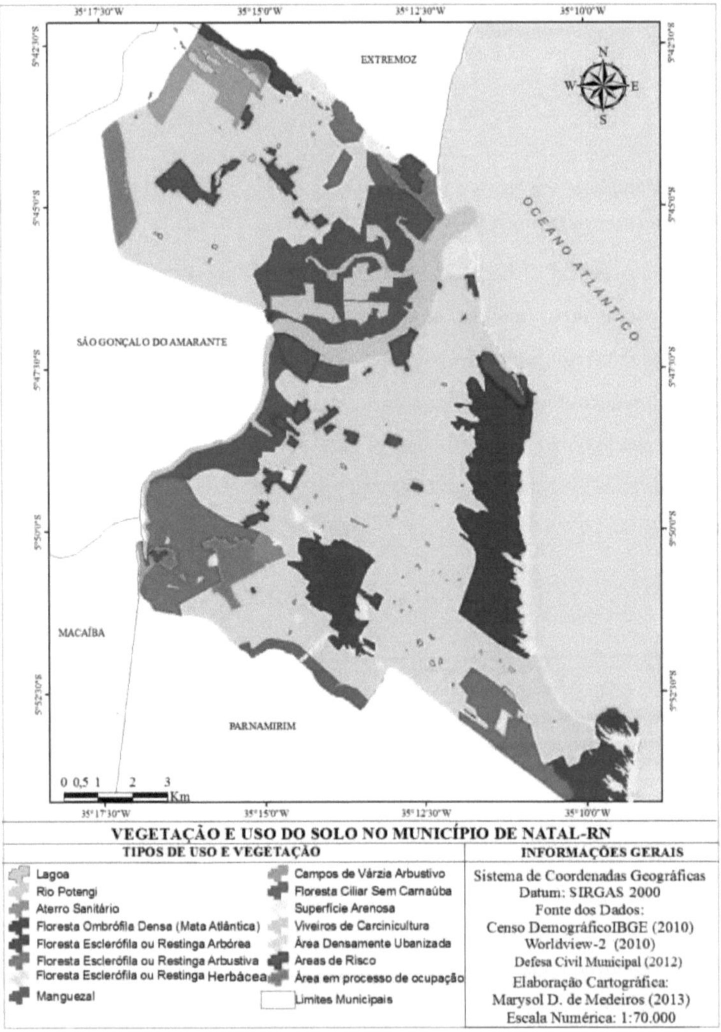

Figure 45: IP Vegetation and Land Use. **Source:** Prepared by Marysol Medeiros based on data from IBGE (2010), Municipal Civil Defense (2012), *Worldview-2* images (2010).

The original methodology did not include urban areas, however, given that the capital of Rio Grande do Sul is very densely populated, it was necessary to adapt it so that this very significant element was taken into account. To do this, it was only necessary to vectorize the areas of urban occupation according to their stage of densification.

Man's actions affect the landscape and its natural dynamics, so all land uses were classified as having very high vulnerability, with only the value varying. The extreme value of vulnerability, equivalent to 3.0, is found in the risk areas mapped by the Municipal Civil Defence, which have poor urbanization standards and precarious housing, which in turn is located in areas of environmental risk. Also due to the lack of adequate urban infrastructure, the areas in the process of occupation were classified with vulnerability equal to 3.0, since they are areas without paving, drainage, public lighting and other urban facilities, which means that their absence causes various inconveniences for the population living there. The area of the old Cidade Nova rubbish dump was also given the same vulnerability rating due to the high risk of contamination it poses, as well as the environmental impact it has had for years and has still not ceased to have. The densely urbanized areas had a vulnerability score of 2.9 because, despite the high level of soil sealing, they have portions with adequate urban infrastructure. Shrimp ponds, on the other hand, received a score equivalent to 2.8 because this is an activity that greatly affects the development of mangroves and can cause contamination of the estuary's water and consequently of the entire food chain that lives there.

The lagoons were also classified as having very high vulnerability, with a value of 3.0, given that most of them are located in areas of high urban density and, moreover, they easily overflow, affecting all the buildings that surround them. Sandy soils, or exposed soils, were also classified in this way due to their susceptibility to triggering morphogenesis, as well as areas with herbaceous restinga, since this is vegetation that develops on soils with strong sedimentation and little stability.

The density of the vegetation cover is the main factor in establishing the vulnerability classes. The shrubby restinga was thus classified with a high Vulnerability value of 2.6, as its size does not guarantee the density needed to completely protect the soil from weathering agents.

With values equivalent to 1.7 and consequently showing low vulnerability due to the density of the vegetation cover are the shrubby grasslands, the shrubby restinga, the mangrove forest and the riparian forest.

Due to the high density of the vegetation, which protects the soil on which it rests in a healthy way, the Dense Ombrophilous Forest was classified as having very low vulnerability and received a value equivalent to 1.0.

Table 13: Summary of vegetation cover and land use and respective vulnerability values

Types of use/vegetation	**Main features**	**Weight**
Lowland Dense Ombrophylous Forest	Atlantic forest. With a high density of vegetation that protects the soil.	1,0
Shrubby Varzea fields	They are found along rivers and lakes, with the presence of fresh water flooding the surrounding soils, leaving them wet, with a diverse physiognomy, sometimes sparse, sometimes dense, and a height of between 0.3 metres and 2 metres.	1,7
Riparian forest without carnauba	It is found along rivers and occupies a strip no longer than a few tens of meters; in some stretches it is dense and exuberant, perennial.	1,7
Mangrove Forest	It is found in the estuary of the Potengi River, with a uniform physiognomy and height varying between 6 and 10m, perennial. It is currently very degraded.	1,7
Sclerophyll Forest or Restinga Arbòrea	It is easily found anchoring the city's palaeodunes. Its sclerophyllous leaves fall off during the driest period of the year, leaving the soil more exposed to erosion, but its arboreal size is better able to protect the soil.	1,7
Sclerophyll Forest or Shrubby Restinga	It is easily found clinging to the city's palaeodunes, with sclerophyllous leaves that fall off during the driest period of the year, leaving the soil more exposed to erosion.	2,6
Shrimp Nurseries	It occupies mangrove areas in the Potengi River estuary, where the mangroves have been devastated to make way for shrimp larvae breeding ponds.	2,8
Densely urbanized area	Areas of strong urban pressure, with some areas without adequate planning. Strong soil sealing.	2,9
Lagoons	They arise from the resurgence of the groundwater table, or are artificial to capture rainwater from urban drainage. Most of them are located in areas of high urban density and easily overflow, affecting all the buildings around them.	3,0
Rich areas	Areas without proper planning and lacking in basic infrastructure such as sewage, paving and drainage, as well as	3,0

	precarious housing conditions located in areas of great vulnerability.	
Herbaceous Restinga	The herbaceous restinga is found close to the sea, with a variable density and its main representatives are grasses.	3,0
Area in the process of being occupied	Adjacent to areas of consolidated occupation with reasonable infrastructure or areas of peripheral expansion, far from the urbanized core, with low density of occupation and lacking basic infrastructure.	3,0
Cidade Nova landfill	It does not have the capacity to receive more waste, it is located in an area of growing urban density and it is located on dunes, where the leachate from the waste is easily absorbed by the soil and contaminates the aquifer.	3,0
Sandy surface	They present completely exposed soil and completely vulnerable to weathering.	3,0

Source: Prepared by Marysol Medeiros.

The following figure (45) highlights the vulnerability for each type of vegetation cover, as well as for each land use pattern.

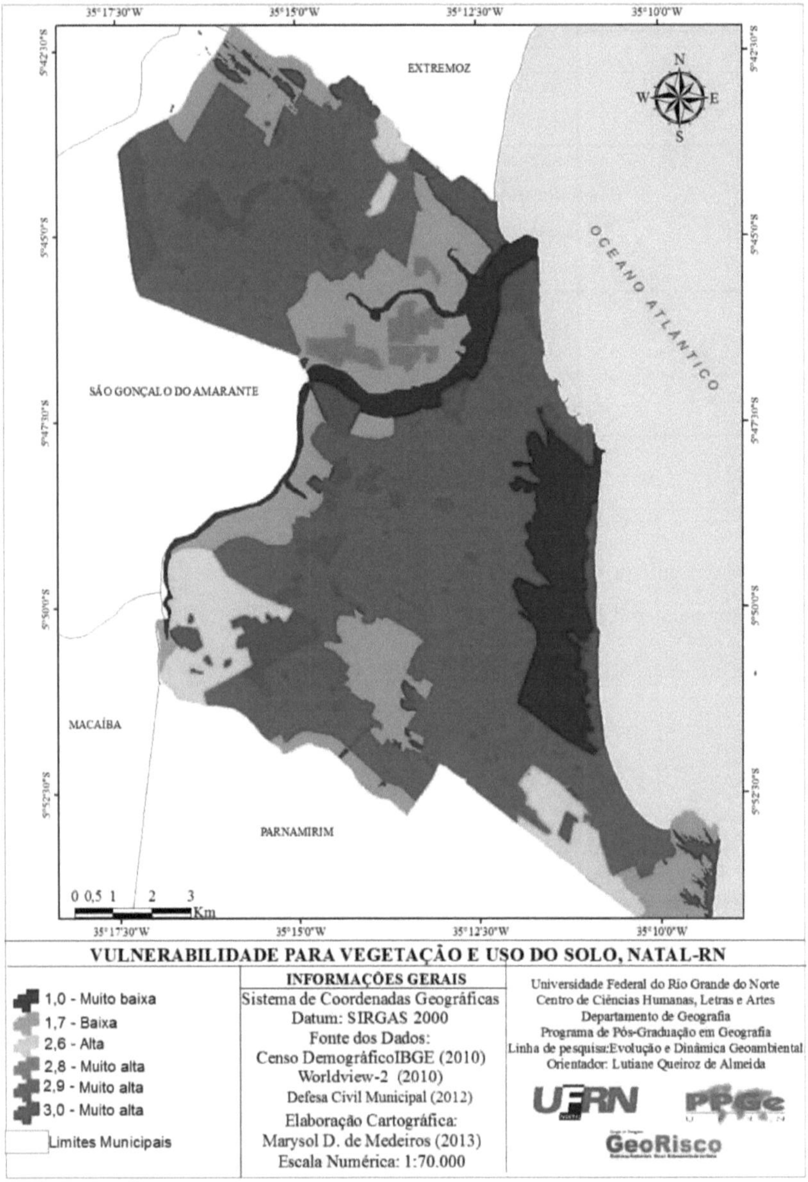

Map 45: Vulnerability to Vegetation and Land Use. **Source:** Prepared by Marysol Medeiros based on data from IBGE (2010), Municipal Civil Defense (2012), Worldview-2 images (2010).

4.3 Spatialization of Natal's Environmental Vulnerability Index

The Environmental Vulnerability Index was drawn up by overlaying the thematic maps drawn up earlier with the Information Plans for each component of the landscape. The *Field Calculator* tool in the Arcgis 9.3 GIS program was used to calculate the arithmetic mean of the intersection of the polygons containing data from the five Information Plans.

From the analysis of the Environmental Vulnerability map (figure 50), it can be seen that the environmental vulnerability values for the city range from medium to very high, with the areas classified as having medium vulnerability being those where there is little or no urban occupation.

Despite being somewhat alarming, Natal's environmental vulnerability is still high, given that the substrate on which the city has developed is basically made up of unconsolidated material that is easily transported, especially water. On the one hand, the sandy soil facilitates the infiltration of water, preventing surface accumulation; however, the urban growth of the city has waterproofed areas that were once covered by dunes; not forgetting to mention that, in order to make way for urban expansion, several natural lagoons were filled in and, as they constitute land with lower altimetry levels, the buildings arranged on these former lagoons are constantly flooded when there is more intense rainfall. Neighborhoods like Lagoa Seca and Lagoa Nova were named after the lagoons they contained. The State Administrative Center, for example, was built over landfilled lagoons and in rainy seasons the consequences can be flooding, as shown in figure 46.

Figure 46: RN Administrative Center after rains in the 1990s. **Source:** Google Images.

The highest Vulnerability values are found on the slopes of the dunes, due not only to the type of material that makes up these features, but also to the high degree of slope they present, making them extremely unstable environments and forming areas vulnerable to landslides of friable material. Many homes are found in this risky situation, especially in the neighborhoods of Cidade Nova, Planalto, Felipe Camarao, Guarapes and Mae Luiza, because they are located on the slopes of the dunes, and in most cases the vegetation has been removed to make way for these irregular constructions. The images below illustrate this situation.

Figure 47: Images of occupation in dune areas. Left: street with a "containment" wall (indicated by arrows) to stop the sand sliding off the dune in the Cidade Nova neighborhood. Right: modest houses in the Baixa do Cao community (Cidade Nova) built at the base of the dune. **Source:** author's personal archive, September 2011.

Figure 48: Images of occupation in dune areas. Left: irregular occupation of the dunes of Parque da Cidade in the Mae Luiza neighborhood. Right: houses occupying dune slopes in the Guarapes neighborhood. **Source:** author's personal archive, May 2013.

Deslizamento de duna derruba muro de casa

Natal tem doze áreas com risco de desabamento

TRIBUNA DO NORTE

Notícias Esportes Blogs&Colunas VcNotícia Cultura&Lazer

Plantão Natal Política Economia Nacional e Internacional Multimídia Cinema Copa 2014

Natal

Erosão deixa 21 cidades sem energia

Recomendar 12 pessoas recomendaram isso. Seja o primeiro entre seus amigos. Tweet

O tombamento de estrutura em uma linha de transmissão da Companhia Energética do RN interrompeu o fornecimento de energia elétrica, por volta das 12h desta quarta-feira, em 21 cidades da região Agreste do Estado - sobretudo nos município localizados ao Sul da capital potiguar. De acordo com informações repassadas pela assessoria de imprensa da Cosern, a queda do equipamento foi provocada pela erosão em área de duna localizada no bairro do Planalto, zona Oeste de Natal.

Várias equipes da empresa foram mobilizadas para trabalhar no local, no intuito de agilizar o restabelecimento do fornecimento, e segundo a assessoria da Companhia a situação teria sido normalizada às 20h15.

O problema ocasionado devido as chuvas intensas que assolam Natal, atingiu cerca de 170 mil pessoas nos municípios de Arêz, Baía Formosa, Brejinho, Canguaretama, Goianinha, Lagoa Dantas, Lagoa Salgada, Macaíba, Montanhas, Monte Alegre, Nísia Floresta, Parnamirim, Passa e Fica, Santo Antônio do Salto da Onça, São José de Mipibu, Serrinha, Tibau do Sul, Vera Cruz e Vila Flor. Alguns bairros de Natal e parte do município de Lagoa de Pedra também tiveram o fornecimento de energia prejudicado.

Figure 30: News taken from the *online* newspaper Tribuna do Norte referring to events linked to dune landslides. On the left: news of the collapse of a house in the Cidade Nova neighborhood, Natal/RN. Right: Toppling of a power transmission structure after erosion of a dune in the Planalto neighborhood. **Source:** http://tribunadonorte.com.br/. Published on October 22, 2010/July 5, 2012.

The areas with high vulnerability largely correspond to the city's urban areas, but this does not mean that they are unsuitable for occupation. However, when drawing up the Vulnerability map for the Vegetation and Land Use PI, this spatial section received a vulnerability value of 2.9, due to the heterogeneity of the city's regions, with areas that have urban facilities and many without any urban infrastructure.

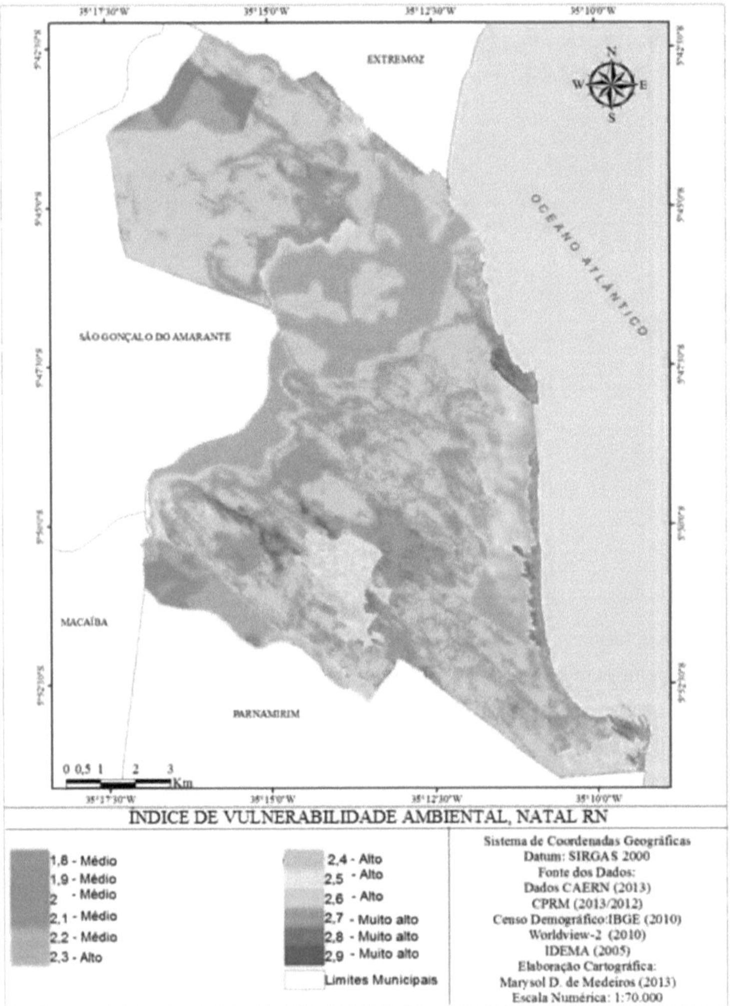

Figure 50: Environmental Vulnerability. **Source:** Prepared by Marysol Medeiros based on data from IBGE (2010), CAERN (2013); CPRM (2012/2013) Municipal Civil Defense (2012), *Worldview-2* images (2010) and IDEMA (2005).

Allied to this factor, the urban soil is extremely impermeable and, in the case of the capital of Rio Grande do Norte, there are few streets with rainwater drainage systems, and many of these are blocked by the accumulation of garbage or clandestine sewage connections.

The number of lagoons in the city and the low altitudes mean that there are several flooding points. Allied to these factors, the majority of the city's lagoons are intermittent, which means that many are landfilled during dry periods to expand spaces for construction, greatly increasing the accumulation of surface water.

Despite the sandy soil, drainage studies for the city show heavy rainfall equivalent to 98.4 mm in 24 hours, as can be seen in the following table from the Risk Reduction Plan for Natal. This data, combined with the infiltration rates for the different types of soil in the city, shows the parameters needed to understand that Natal has areas of flooding, but that, depending on the type of soil, the accumulation of surface water lasts only a few hours, which is enough time for residents whose homes are invaded by water to suffer losses.

Table 14: Intense Rainfall for the city of Natal, from the equation adjusted by Otto Pfafstetter, 1982

Duration		min	min	min	time	hours	hours	hours	hours	hours	hours	days	days	days
		5	15	30	1	2	4	8	14	24	48	3	4	6
Recurrence (years)	2.33	10,2	20,0	28,8	38,5	48,1	58,7	71,0	83,2	98,4	127,7	152,9	176.8	222,7
	10	11,4	23,9	37,0	52,0	65,9	81,4	98,7	115,3	135,6	175,0	207.8	238.8	299,0
	25	12.4	26,8	42,5	61,1	78,1	97,2	118,0	137,6	161,2	207,3	244.8	280.3	349,7
	50	13.4	29,1	46,9	68,3	87,9	110,0	133,9	155,8	182,1	233,4	274,5	313.5	390.0

Source: Acquatool (2008)

Table 15: Infiltration time of heavy rainfall according to the city's soil pattern

Types of soil	Location	Rate of Infiltration mm/h	Total infiltration in heavy rain
The sandy soils of the preserved dune covers	Land in the Dunes and City parks and the Lagoinha dune cordon	25	On average 4 hours
Uncharacterized dune cover	Neighborhoods of Candelaria and Mirassol	13	On average 8 hours
Uncharacterized dune cover	Lagoa Nova, Dix-Sept Rosado, Tirol and Petrópolis and in the North Zone	7	14 hours on average
Formations of lagoon bottom soils, alluvial soils and Barreiras outcrops	Capim Macio, Ribeira and Barreiras outcrops	3	On average 33 hours

Source: Prepared by Marysol Medeiros based on data from the Drainage and Stormwater Management Master Plan for Natal (2009).

Based on this data, the following table was drawn up with the monthly frequency of maximum rainfall (equal to or greater than 60 mm) in 24 hours in Natal/RN between the years 1992 and 2011 and with this it can be seen which months have seen flooding in the city.

Table 16: Monthly frequency of rainfall above 60mm/day in Natal between 1992 and 2011

Event in	Months of the year											
	Jan	Feb	Sea	Apr	May	Jun	Jul	Aug	Set	Out	Nov	Ten
60 - 80	2	4	6	10	6	15	8	2	1	0	0	0
80 -100	2	0	5	0	1	7	2	1	0	0	0	0
100 -120	2	0	2	3	1	3	1	1	0	0	0	0
120	0	0	0	1	3	4	5	1	0	0	0	0
Total	6	4	13	14	11	29	16	5	1	0	0	0

Source: Prepared by Marysol Medeiros from EMPARN data.

As you can see, the months with the most heavy rainfall are June and July, with days of rainfall of more than 120 mm, leading to the spread of more news like the following:

Figura 51: News taken from Tribuna do Norte about the disruption caused in the city by the flooding.
Source: http://tribunadonorte.com.br.

Intense rainfall data is used to guide actions to combat flooding and landslides, such as those that occurred in Mae Luiza, the Jacó Community and Passo da Pàtria. So, knowing that the city has 98.4mm/h as heavy rain, a 60mm rainfall in 8 hours is enough to put bodies like Civil Defense on alert.

Vulnerability only exists if someone or something is exposed to some kind of potentially harmful event. In this way, the social vulnerability index was developed to identify the places in the city of Natal where the population is potentially most vulnerable to environmental risks. Meanwhile, it was necessary to characterize the natural aspects of the city that make certain areas more vulnerable to environmental risks, in this case flooding and landslides.

However, the overlapping of the Social Vulnerability Index and the Environmental Vulnerability Index showed where in the city the areas of Socio-Environmental Vulnerability are found in the Potiguar capital.

In this way, the following section will discuss the development of the Socio-environmental Vulnerability Index for the city of Natal, showing how this index was developed and the results obtained through its application.

5. SOCIO-ENVIRONMENTAL VULNERABILITY IN NATAL

Measuring Socio-environmental Vulnerability is complex, as it deals with different spheres that are at the same time intrinsic to each other: the social and the environmental; since the phenomena that occur in the environment reflect directly on society, and society exerts strong pressure on the environment.

Both society and nature have dynamic, wide-ranging and diverse characteristics, while proposing the measurement of social and natural conditions requires as much information as possible about the area to be studied, since the greater the volume of data, the more nuances can be captured.

Even with all the data, exhaustive mapping and studies, some facets of vulnerability cannot always be grasped, because there are particular characteristics, inherent in the individual's behavior, that can make them more vulnerable, an example of which is the difficulty of measuring the perception of risk.

However, the elaboration of Vulnerability Indices shows in a synthetic way the main attributes responsible for the Vulnerability of an area or a group in the face of danger. In order to do this, Vulnerability Indices must answer questions such as: Who is Vulnerable? What risks are they Vulnerable to? Who is most vulnerable? Where are they most vulnerable? Why are they more Vulnerable?

In this way, the socio-environmental vulnerability index provides a cross-section of the social and environmental factors responsible for an area's greater or lesser exposure to natural risks, which are aggravated by the lack of preparation or conditions for individuals to withstand and deal with danger.

5.1 Methodology for drawing up the Socio-environmental Vulnerability Index

The methodology used in this stage of the work consists not only of overlaying social and natural data, but above all of crossing them through a correlation matrix, which can be seen in figure 52.

VULNERABILITY INDEX SOCIAL (IVS)		VULNERABILITY INDEX ENVIRONMENTAL (VAT)
1. Very high	4. Medium Low	A. Very High
2. High	5. Low	B. Alta
3. Medium High	6. Very low	C. Average

SOCIOENVIRONMENTAL VULNERABILITY INDEX (IVSA) = (IVS x IVA)

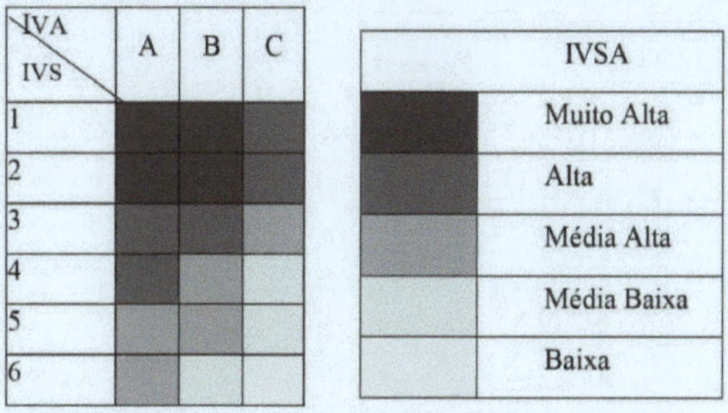

Figura 52: Methodology for drawing up the legend for the Socio-environmental Vulnerability map.

Source: Prepared by Marysol Medeiros.

As shown in the previous chapters, the Social Vulnerability Index (SVI) and the Environmental Vulnerability Index (EVI) were organized separately. The methodology used to prepare the SVI analyzes social variables listed by the IBGE demographic census that denote social disadvantage; these variables were compiled resulting in social vulnerability factors, with a spatial scale at the level of the census sector. To prepare the IVA, environmental factors that favor morphogenetic processes were classified as high vulnerability, while factors in which pedogenetic processes prevail were classified as low environmental vulnerability. However, as this is a methodology that is directly related to soil loss, it was necessary to take into account the topography and layout of the city's water bodies in order to diagnose the areas prone to flooding.

For the iVS, six vulnerability classes were listed: very high, high, medium high, medium low, low and very low. However, the city of Natal did not have any areas with very low or low environmental vulnerability, so it was necessary to cross-reference the six classes of social vulnerability with three classes of the environmental vulnerability index, which are: medium vulnerability, high vulnerability and very high vulnerability.

Using the *Field Calculator* tool in the ArcGis 9.3 program and the principles of Boolean logic[8] with the use of the AND operation, it was possible to perform the algebra of the thematic maps with the crossing of the polygons. Image 53 shows the above procedure in the geographic information processing program.

---

8 Boolean logic, or Boolean algebra, is a system of mathematical logic created by the English mathematician George Boole (hence the name Boolean algebra). It is used to create logical rules or expressions used to analyze, select and process data.

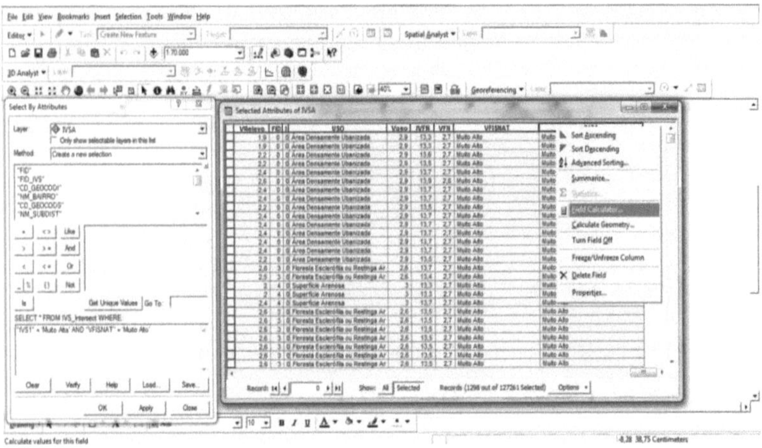

Figure 52: Screen shot of the ArcGis 9.3 GIS program showing how the map algebra was carried out to produce the IVSA. **Source:** Personal archive.

5.2Spatialization of the Socio-environmental Vulnerability Index

Overlaying the social and environmental vulnerability indices resulted in the socio-environmental vulnerability index. This methodology produced satisfactory results, given that many areas of the city with high socio-environmental vulnerability were covered.

As can be seen in the socio-environmental vulnerability map (figure 59), several areas of the city already known for having recurrent episodes of flooding, landslides or flooding have obtained high socio-environmental vulnerability values. However, areas lacking in urban infrastructure and with a poorer population also had high socio-environmental vulnerability values.

The main areas of socio-environmental vulnerability are located mainly in the north and west of the city of Natal, as these are geo-environmental features that are conducive to the occurrence of natural hazards.

In these areas of the city, it is common to find irregular occupations on dune slopes, which have high altitudes and slopes, causing the friable material (sands) to slide due to eolian transport in dry periods, often culminating in houses being buried. During periods of more intense rainfall (in the city of Natal between April and July), due to the saturation of the soil by the high amount of water in the system, together with inadequate soil management (with irregular cuts in the land, structures without adequate foundations and removal of the vegetation cover), deep cracks in the walls of homes are common, greatly compromising the safety of the buildings and often causing landslides, as can be seen in the images below.

Figure 54: Images of occupation of slopes. Left: rubble left over from a house that collapsed after heavy rains in 2013, in the Sao José do Jacò community, in the east of the city. Right: house "invaded" by dune sand in the Cidade Nova neighborhood. **Source:** author's personal archive, September 2011/November 2013.

In areas of low topography, such as the banks of rivers and lagoons, flooding is common during periods of heavy rain. The banks of the Rio Doce in the north of the city have been occupied more intensively and the removal of the riparian forest has silted up the river, making it shallower and, consequently, when the volume of water in the system increases, all the surrounding buildings are flooded. The same happens in the Passo da Pàtria community on the east side of the city, where the proximity of buildings and the accumulation of garbage and debris in the Baldo Canal leads to flooding. The Riacho das Lavadeiras on the west side has been completely canalized, however, the low topography of the site and the sealing of the soil in the areas surrounding the creek is conducive to the concentration of rainwater, so it is very likely that the canal will not be able to withstand the large volume of water. The images below illustrate the above.

Figure 55: Images of occupation on river banks in the city of Natal. Left: debris and garbage accumulated in the Baldo Canal, making it silted up, and constructions near the canal making this area highly vulnerable to flooding. Center: removal of riparian forest and construction along the Rio Doce, in the north of Natal, Right: Riacho das Lavadeiras, Quintas district, west side, is canalized, but the low topography and sealing of the soil in the areas surrounding the stream is favorable to the concentration of rainwater. **Source:** author's personal archive, September 2011/November 2013.

The areas around natural lagoons or large artificial lagoons (catchment lagoons built for the convergence of rainwater that runs off the surface) also had high socio-environmental vulnerability, especially the areas near the lagoons in the neighborhoods of Neópolis, Nova Descoberta, Ponta Negra and Cidade da Esperança due to the high degree of urban occupation around them.

Figure 56: Images of stormwater drainage catchment ponds. Left: catchment pond on Avenida da Integraçâo on a rainy day. You can see that it is about to overflow. Right: Lagoa de capitaçao in the Capim Macio neighborhood. **Source:** author's personal archive, August/November 2013.

Because it was built on dunes, the topography of the Potiguar capital has slopes that favor the accumulation of surface water, since the portions with the lowest altimetry levels form veritable lagoons during rainy periods. Flooding spots can be found in practically every part of the city.

Figure 57: Image of the Escadaria Community, Praia do Meio. As can be seen, the topography of the street as well as the sealing of the soil favors the descent of the water that runs off the surface. **Source:** author's personal archive, November 2013.

The mangrove swamp that covers the Potengi River estuary is also the target of urban sprawl, in which this geomorphological feature has been inappropriately filled in and where it is sometimes possible to detect the presence of garbage as the material used. Buildings in these conditions are subject to ground settling and, as a result, can develop cracks that compromise the safety of the building.

Figure 58: Image of the Mosquito Community, Nordeste neighborhood. Houses built inside the mangrove on landfill made with garbage. **Source:** author's personal archive, September 2011.

As can be seen on map 23, the city of Natal has large areas of socio-environmental vulnerability, totaling 12 points of very high vulnerability. The following table summarizes which risk areas in the city have very high socio-environmental vulnerability, describing the types of risk and a brief description of the area.

Table 09: Areas of very high socio-environmental vulnerability in Natal

	Neighborhood/Location	Type of risk	Description
1	Monte Celeste/ Plateau	Erosion, waste disposal, flooding.	A community with precarious housing near the Cidade Nova garbage dump.
2	High Voltage / Guarapes	Landslides and flooding.	It is located in a high-voltage strip of land, irregularly occupied by dunes and close to the old garbage dump.
3	Leningrad	Landslides and flooding	It is located in a high-voltage strip of land, irregularly occupied by dunes and close to the old garbage dump.
4	Settlement Anatália de Sousa	Duneslide	Occupation of dune strip by rammed earth houses.
5	Guarapes / banks of the Potengi-Jundiai estuary	Flooding and high risk of landslides	Precarious housing located between the banks of the Potengi/Jundiai river estuary and the slopes of the ZPA-4 dunes on the banks of the BR 226 highway.
6	Baixa do Cao/ Cidade Nova	Landslides and flooding	Irregular occupation of the dunes in ZPA-01.
7	Felipe Camarao between Av. Perimetral Leste, Rua dos Campos and Av. Capitao Mor Gouveia	Flooding, debris flows, landslides and collapsed barriers.	A low-income community is located in a high-voltage power strip.
8	Behind the Bom Pastor Cemetery	Garbage loading, dune encroachment and landslides.	Low-income residents occupying dune areas.
9	Redinha Near Nilton Navarro Bridge	Flooding	Unpaved roads with garbage piles occupying mangrove and dune areas.
10	Conjunto Jardim Progresso/ Nossa Senhora da Apresentaçao	Several flooding points.	Households in areas without adequate urban infrastructure and economically deprived populations.
11	Pajuçara/ Genipabu Road	Landslide and flooding of the Rio Doce	Households occupying dune areas and close to the Rio Doce.
12	Community of Africa/ Redinha	Landslides	A community of precarious dwellings occupying mobile dunes.

Source: prepared by Marysol Medeiros based on survey data.

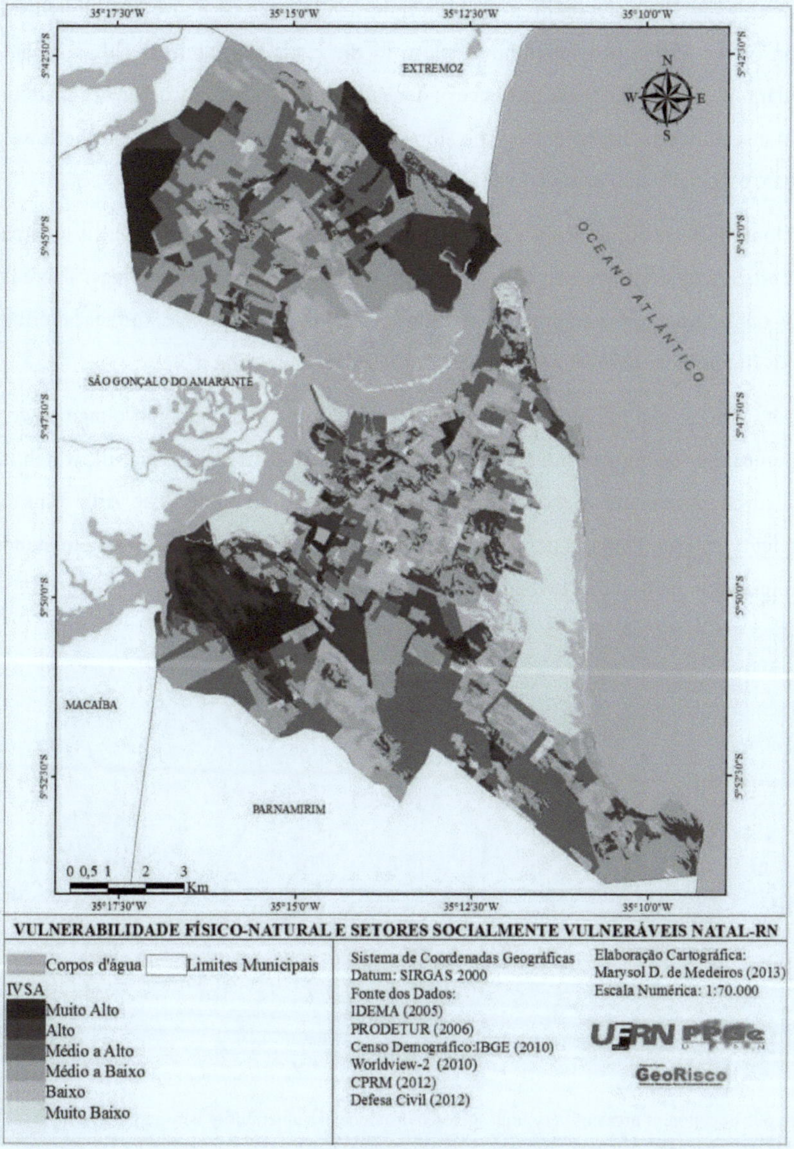

Map 59: Socio-environmental vulnerability. **Source:** Prepared by Marysol Medeiros based on data from IDEMA (2005), PRODETUR (2006), IBGE (2010), *Worldview-2* images (2010), CPRM (2012) and Municipal Civil Defense (2012).

The satellite images show the location of some of the points mentioned in the table above. The first five points are located in the west, between the Planalto and Guarapes neighborhoods. The polygon in the bottom right-hand corner of figure 60 highlights a portion of the Planalto neighborhood, close to the city's old landfill. This area has irregular topography due to the presence of dunes, with soils sealed by urban occupation, and is susceptible to landslides on the slopes of the dunes and flooding in the lower portions, especially on Paracati Avenue.

The Monte Celeste community, as has already been discussed in this paper, is an area in the process of slumification, with a population that has many shortcomings. There is a gully on the edge of the community with an accumulation of garbage, further favoring the movement of material from the edges of the gully, increasing it.

The Alta Tensao and Leningrado communities are located under a high-voltage power line (hence the toponymy) and are also close to the old Cidade Nova landfill, which is occupied by dunes and is susceptible to landslides and flooding. The Alta Tensao community and the Planalto neighborhood should belong to the isolation ring to avoid possible contamination due to their proximity to the old landfill.

Figure 60: Location of areas of very high socio-environmental vulnerability between the Planalto and Guarapes neighborhoods. Monte Celeste community; 2. Alta Tensao community; 3. Cidade Nova garbage dump; 4. Leningrad and 5. Planalto. **Source:** Google Earth, March 2014.

In the Guarapes neighborhood, there is a settlement known as Anatalia de Souza, which occupies an area adjacent to dunes. Most of the houses in this settlement are made of rammed earth, but there are also shacks near vegetated dunes, which present a medium risk of landslides.

On the banks of the Potengi/Jundiai river estuary, there are houses that are susceptible to flooding (see image 61), as the homes occupy mangrove areas that are subject to constant flooding. Crossing the BR-226 there are houses occupying slopes of the dunes that make up ZPA-04 with a high risk of landslides, not to mention the proximity of the homes, both those in mangrove areas and those in the dunes, to the BR-226 (which divides these two areas), as they are exposed to risks involving traffic accidents.

Figure 61: Location of the areas of socio-environmental vulnerability in Guarapes. 1: Anatàlia de Sousa settlement; 2: occupation of the estuary. **Source:** Google Earth, March 2014.

The Cidade Nova neighborhood is made up of several vegetated dunes that are heavily pressured by intense urbanization. One area with recurring social problems and high natural risks is the Baixa do Cao community, which, because it occupies the slopes of the dunes on the edge of ZPA-01 (figure 41), faces serious problems of landslides, causing the population to improvise ways of containing the descent of the sand (figure 62). Another problem stems from the low topography and waterproofing of the soil, which means that flooding occurs during periods of heavy rainfall.

Figure 62: Occupation of dune slopes in the Cidade Nova neighborhood. 1) Baixa do Cao **Source:** Google Earth, March 2014.

Figure 63: Images of the Baixa do Cao/ Cidade Nova community. Left: improvised engineering with tires to contain the descent of material from the dune. Right: house with a brick bar at the entrance to prevent water from entering when the area floods and a wall showing the level the water reached during the last flood. **Source:** author's personal archive, September 2011.

The main causes of socio-environmental vulnerability in the city are directly linked to a lack of planning, with many streets being heavily waterproofed and steeply sloping, favoring the faster and more concentrated flow of surface *water* (*runoff*), Not only that, but the lack of sanitation and awareness among residents is responsible for the discharge of wastewater from homes into the streets, which flows randomly throughout the day, favoring laminar erosion as well as being a constant source of disease transmission. Another very obvious factor in the city is the low number of streets that have a rainwater drainage system. Allied to this, the accumulation of garbage on public roads leads to the obstruction of culverts and manholes, hindering rainwater drainage and directly contributing to flooding, as well as the carrying of this garbage, which leads to erosion.

The sectors with very high and high social vulnerability continued to have high values for socio-environmental vulnerability. Thus, sectors in the neighborhoods of Guarapes, Planalto, Cidade Nova and Felipe Camarao, in the West zone; the neighborhoods of Candelaria (Viaduto Community), Neópolis and Nova Descoberta, in the South zone; Alecrim in the East zone and in the North zone in the neighborhoods of Redinha, Pajuçara, Lagoa Azul and Nossa Senhora da Apresentaçao, accounted for very high socio-environmental vulnerability.

Sectors with medium-high or medium-low social vulnerability, when superimposed on the environmental vulnerability map, showed high socio-environmental vulnerability, as is the case with the sectors that make up the Sao José do Jacó Community in the Rocas neighborhood and the Mae Luiza neighborhood. These areas have poorer individuals in terms of income, but the standard of construction (masonry houses), paved streets, public lighting and rainwater drainage systems mean that the social vulnerability index in these areas is not so high. However, residents are constantly surprised by the imminent threat of landslides.

It is not common in the capital of Rio de Janeiro for major natural disasters to have fatalities; in most cases, material losses are reported. However, the apparent tranquility in the face of environmental vulnerability should not be viewed in an apathetic manner by Civil Defense managers and agencies, since the news often shows cases of cave-ins and house collapses, as the newspaper clipping below shows.

the Tribuna do Norte electronic newspaper alluding to the house that

Parte de residência em área de risco desaba na zona Leste de Natal

Publicação: 03 de Julho de 2013 às 11:03 | Comentários: 0

As chuvas de ontem e madrugada desta quarta-feira (3) provocaram o desabamento de parte de uma residência na comunidade do Jacó, no bairro das Rocas, zona Leste de Natal. A área tem residências em situação de alerta e muitos moradores relutam em deixar os imóveis. O proprietário da casa que teve parte da estrutura derrubada foi um dos que resistiu até ser convencido a deixar o local.

Figure 64: News removes Rocas neighborhood (Sao José do Jacó Community) after heavy rain in the city. **Source:** http://tribunadonorte.com.br/. Published on July 3, 2013.

In areas with residents with a higher economic profile, such as the neighborhoods of Capim Macio, Ponta Negra, Lagoa Nova and San Vale (in the neighborhood of Cidade Satélite), socio-environmental vulnerability became high due to the lack of urban equipment and services, such as deficiencies in the drainage system, the accumulation of garbage in the streets and the lack of paving, making them flooding hotspots every year. Gender was another relevant social vulnerability factor for the high socio-environmental vulnerability index, given that these sectors had more homes with no male residents.

These events do not result in human losses, but the material losses are countless, not to mention the inconvenience faced by the population, whose access roads are blocked by the accumulation of water.

Figure 65: News taken from the Tribuna do Norte online newspaper referring to areas of flooding in the south of the city. **Source:** http://tribunadonorte.com.br/. Published on June 28, 2012.

Field visits between 2011 and 2014 showed that the situation in these areas and many others in the city has not changed. The city's Civil Defense department advises residents in areas of risk to leave their homes, but there is a lot of resistance. Despite being warned of the risks they run, they don't want to leave their homes because, according to most of the residents, they have no other option or have strong emotional ties to the place.

Figure 66: Images of houses with compromised structures and residents. On the left: cracks in the structure of a house in the Sao José do Jacó community in the Rocas neighborhood. On the right: a house

with a wall that prevents the "invasion" of the dune showing cracks and embrittlement in the Mae Luiza neighborhood, not just the wall, but the whole house is in this state. **Source:** author's personal archive, May 2013.

Measuring socio-environmental vulnerability is an arduous task given the difficulties encountered in obtaining data capable of characterizing reality as accurately as possible.

Not only the selection of data, but also its processing requires a series of choices that can only be made if there is prior knowledge of the area to be studied. Nevertheless, the development of a socio-environmental vulnerability index is an important tool for informing public management decisions in a municipality, for example.

FINAL CONSIDERATIONS

The city of Natal has several risk areas, mainly on the outskirts of the city, due to the occupation of spaces that have legal and/or natural limitations which, combined with the lack of urban organization, increase the vulnerability of the population living in these areas.

The importance of studies that map the risk areas of a given territory is due to the fact that prevention measures become more effective, since it is easier to see the areas of vulnerability due to the spatialization of these risks.

Work of this type serves as a diagnosis capable of identifying the type of risk, the spatialization and the degree of vulnerability of large areas, making it possible to obtain the information necessary for planning risk areas, answering important questions such as: What is the risk? Who is vulnerable? What is the degree of vulnerability?

In order to meet the main objective of this research, which was to map areas of social vulnerability and environmental risks in the city of Natal, it was necessary to take into account the interrelationships between social vulnerabilities and differentiated exposure to natural risks. To this end, it was necessary to draw up indices of social and environmental vulnerability.

Each chapter of this work was designed to meet a specific research objective, so the first chapter discussed the concepts and theoretical and methodological approaches to risk and vulnerability from the perspective of Geography.

It was necessary to discuss the definitions of risk, danger and vulnerability, since they are treated synonymously by common sense, but these are important secondary concepts for studies involving risks and vulnerabilities in the field of Geography.

Vulnerability studies are extremely complex, firstly because the concept is used by various sciences, making it difficult to choose a definition that is universal for all areas of knowledge. Secondly, the measurement of vulnerability requires extra care, given the multiple possibilities of methodologies, as well as the difficulty of gathering data to characterize reality as reliably as possible. Thirdly, it is difficult to grasp an individual's or group's perception of risk due to its great subjectivity.

The need to explain the definitions of the concepts of risk, danger and vulnerability arose because these three are intrinsically linked, since there is only risk if there is danger and the degree of vulnerability of the individuals exposed.

The second chapter discusses the social vulnerabilities that contribute to the formation of risk territories in Natal. In addition, this chapter presents the Social Vulnerability Index (SVI), based on the compilation of data by census sectors from the 2010 IBGE Demographic Census, through a multifactorial analysis of this data.

The Social Vulnerability Index has some nuances to consider. Firstly, the classification with the degree of vulnerability for each sector results in a comparison between the sectors and not in an absolute value. For example, a census tract has an average of 350 households. In the "Improvised Households" variable, a tract in the Alecrim neighbourhood showed very high social vulnerability. However, when analyzing the vulnerability map for this variable in isolation, it is possible to see that out of a universe of 350, only 09 households are improvised, which is not really a value that can significantly represent the sector in general with such a small number. The other sectors, on the other hand, have no improvised homes or a much smaller number, so that when the map was drawn up, this sector was classified as having a high degree of vulnerability.

Sectors such as those found in the Mae Luiza or Rocas neighborhoods did not have such a high degree of social vulnerability due to the infrastructural conditions of some of the streets in these neighborhoods. Despite the notorious poverty of the residents of these areas, with individuals living in unhealthy conditions, they are areas of old urban occupation, where the public authorities *have already* acted to install some urban facilities, with the implementation of some paved roads, with regular garbage collection, public lighting, urban drainage and sewage.

The boundaries of the census sectors do not necessarily coincide with the boundaries of the most deprived communities. This, combined with the fact that the city has contiguous areas of profound social inequality, has meant that, in the same census tract, there are households of low social vulnerability with households of high vulnerability which, when put together, reflect a pattern of average social vulnerability.

There are census tracts close to the mangrove swamp, more precisely in the Felipe Camarao neighborhood, with very low social vulnerability. This is because they are unoccupied areas, just like Parque das Dunas.

It is known that in order to determine social vulnerability it is necessary to investigate variables that go beyond those covered in this work, such as rates of violence, drug trafficking, unemployment and many others, but the lack of data at the census sector level has prevented other variables from being included in the Social Vulnerability Index. The type of construction material used in housing is also extremely important, as it can increase exposure to natural hazards.

In general, the city of Natal has few census tracts classified as having very high social vulnerability, totaling only 11 tracts (1%) in the entire city, with a total area of 417,469m .[2]

The third chapter provides a diagnosis of the environmental factors that contribute to the occurrence of natural hazards in Natal. To this end, various primary and secondary data were compiled for analysis and, subsequently, for the production of thematic maps (geological, pedological, slope and altimetry, vegetation and land use maps).

Adaptations to the methodology were necessary to remedy some limitations, especially for the preparation of the Environmental Vulnerability Index, since the spatial cut-out for which it was conceived did not have a prevalence of urban areas, as it was in the Legal Amazon.

The Environmental Vulnerability Index had to be adapted to take account of the area in question. Thus, it was necessary to include Land Use in the Vegetation Information Plan, since this is an area with a high degree of urbanization. However, it would be more appropriate to insert weights for each Information Plan, thus producing a weighted average rather than an arithmetic average of the values. This is because the type of soil, the morphometry of the relief and the pattern of land use are more preponderant factors in increasing environmental vulnerability in the city of Natal than geology and climate, while the latter does not vary so much in the city.

The fourth chapter deals with the results of the overlapping mapping of social and environmental vulnerabilities in Natal; bringing up the Socio-environmental Vulnerability Index (IVSA). This methodology presented satisfactory results, given that many areas of the city with high socio-environmental vulnerability were covered.

The main areas of socio-environmental vulnerability are located mainly in the north and west of the city of Natal, as these are geo-environmental features that are conducive to the occurrence of natural hazards.

In these areas of the city it is common to find irregular occupations on dune slopes, causing the friable material to slide due to eolian transport in dry periods, often culminating in houses being buried. When rainfall is more intense and the soil is saturated with water, it is common for deep cracks to form in the walls of houses and for landslides to occur.

In areas of lowered topography, such as riverbanks and lagoons, flooding is common when there is heavy rainfall. Such as those that occur on the banks of the Rio Doce in the north of the city, in the community of Passo da Patria and the Baldo Canal in the east and near the Riacho das Lavadeiras in the west.

The areas around the natural lagoons or the large artificial lagoons also had high socio-environmental vulnerability, especially the areas near the lagoons in the neighborhoods of Neópolis, Nova Descoberta, Ponta Negra and Cidade da Esperança due to the high degree of urban occupation around them.

The mangrove swamp of the Potengi River estuary has houses built in areas with inappropriate earthworks, and it is possible to detect the presence of garbage as the material used. These conditions allow the land to settle and, as a result, there may be cracks that compromise the safety of the building.

The main causes of socio-environmental vulnerability in the city are directly linked to the lack of urban planning, combined with the natural characteristics and lack of awareness among residents.

This study based its hypothesis on the overlapping of natural risks and social vulnerabilities in the city of Natal. In this way, the questions that initially arose were answered satisfactorily.

It can be seen that there is indeed greater exposure to natural hazards on the part of less privileged social groups. However, the most vulnerable social groups are not necessarily those with the lowest purchasing power; other factors such as gender and access to urban infrastructure are also preponderant. However, social groups with economic deprivation and difficult access to urban facilities and services are more vulnerable. In addition, many areas with a poorer population are also areas with intense natural dynamics.

Except in places where the population is poor, the natural dynamics are intense, but there is some urban infrastructure, the socio-environmental vulnerability was not as high as expected, as is the case of some sectors in the Mae Luiza and Rocas neighborhoods.

The lack of awareness among the population increases the degree of vulnerability of individuals, not only in terms of risk perception (even under threat of risk, residents do not leave their homes) but also in terms of public cleanliness, since many areas of the city have an accumulation of garbage and open sewage, catalyzing the risks of flooding and erosion respectively. Simple measures such as environmental education projects and drills carried out with the Civil Defense and the community could change this situation.

REFERENCES

ALMEIDA, L. Q. de. **Socio-environmental Vulnerabilities of Urban Rivers: Maranguapinho River Basin, Metropolitan Region of Fortaleza, Cearà**.2010. 278 f. Thesis (Doctorate) - Department of the Institute of Geosciences and Exact Sciences, Universidade Estadual Paulista, Rio Claro, 2010.

. Social vulnerability to environmental hazards. **Revista da ANPEGE,** v. 6, p. 151- 176, 2010.

. Social vulnerability of urban rivers in Brazil: a case study. In: National Seminar on Urban Governance and Metropolitan Development, 2010, Natal. **Proceedings**. Natal, 2010. p. 1 - 17.

. Towards a science of risks and vulnerabilities in Geography. **Mercator**, Fortaleza, v. 10, n. 23, p.83-99, dec. 2011.

ALVES, H. P. F. Methodologies for integrating socio-demographic and environmental data to analyze socio-environmental vulnerability in urban areas in the context of climate change. In: HOGAN, D.; MARANDOLA JR, E. (Org.) **Population and climate change**: human dimensions of global environmental change. Campinas: Nùcleo de Estudos de Populaçao - NEPO/Unicamp; Brasilia: UNFPA, 2009, p. 75-105.

ANGELIM, L. A. de A. et al. **Geology and mineral resources of the State of Rio Grande do Norte**

- Scale 1:500.000. Recife: CPRM - Geological Survey of Brazil, 2007. 119 p.

AZEVEDO, Pablo Guimaraes. **Socio-environmental vulnerabilities in the Environmental Protection Zone - 9, Natal/RN**. 2010. 120 f. Dissertation (Master's Degree in Geography) - Geography Course, Department of Postgraduate Studies and Research in Geography, Federal University of Rio Grande do Norte, Natal, 2010.

BERTRAND, G. Paisagem e geografia fisica global: esboço metodològico. **Cadernos de Ciências da Terra**, Sao Paulo, v. 13, p. 1-27, 1972.

. A transversal geography of crossings: the environment through territories and temporalities. Org: Messias Modesto dos Passos. Maringà: Ed. Maçons, 2007.

BLAIKIE, P. M.; et al. **At risk:** natural hazards, people's, vulnerability, and disasters.

London: Routledge, 1994.

BRAZIL. Ministry of Mines and Energy. General Secretariat. RADAMBRASIL Project. **Sheets SB24/25 Jaguaribe/ Natal**. Rio de Janeiro, 1981. 744p.

BRAZIL. Congress. Senate. Constitution (2012). Law No. 12651, of May 25, 2012. **New Forest Code**. Available at: <http://www.planalto.gov.br/ccivil_03/_ato2011-2014/2012/lei/L12651compilado.htm>. Accessed on: December 20, 2013.

CARDONA, O. D. **The notion of disaster risk:** conceptual framework for integrated management. Manizales: AIDB/ IDEA Program on Indicators for Disaster Risk Management, Universidad Nacional de Colombia, 2003.

CASTRO, C. M. de; PEIXOTO, M. N. de O.; RIO, G. A. P. do. Environmental Risks and Geography: Concepts, Approaches and Scales. **Anuârio do Instituto de Geociências - UFRJ**, Rio de Janeiro, v. 28, n. 2, p.11-30, 06 dec. 2005.

CASTRO, S. D. A. De. Riesgos y peligros: una visión desde la Geografia. **Scripta Nova.**

Barcelona, n. 60, p. 1-15, Mar/2000.

CHAMBERS, R. Vulnerability, coping and policy (Editorial Introduction). **IDS Bulletin.** v. 37, n. 4, Sep 2006.

CHARDON, A. C. Crecimiento urbano y riesgos "naturales": Evaluacion final de la vulnerabilidad global en Manizales, Andes de Colombia. **Disasters and Society**. Red de Estudios Sociales En Prevención de Desastres En América Latina. v. 6, n. 9, p.1-33, 1998.

CHRISTOFOLETTI, A. **Modelagem de sistemas ambientais.** Sao Paulo: HUCITEC, 1979

CUTTER, S. L. Vulnerability to environmental hazards. **Progress in Human Geography,** v. 20, n. 4, p. 529-539, 1996.

S. L.; BORUFF, B. J.; SHIRLEY, W. L. Social vulnerability to environmental hazards. **Social Science Quarterly**, 2003, v. 84 (1): p. 242-261.

CREPANI, E. Instituto Nacional de Pesquisas Espaciais (Org.). **Remote Sensing and Geoprocessing Applied to Ecological-Economic Zoning and Territorial Planning**.

Sao José dos Campos: INPE, 2001. 124 p. Available at:
<http://www.dsr.inpe.br/dsr/simeao/Publicacoes/SERGISZEE3.pdf>. Accessed on: July 1, 2011.

Agricultural Research Company. **Accumulated Precipitation**: Rio Grande do

North, years 1984 to 2011. EMPARN. Available at: <http://189.124.135.176/monitoramento/monitoramento. php>. Accessed on: October 2, 2013.

GUERRA, A. T. **Novo Uicionario geológico-geomorfológico**. 6ª ed. Rio de Janeiro: Bertrand Brasil, 2008.

IBGE - Brazilian Institute of Geography and Statistics. Department of Natural Resources and Environmental Studies. **Technical manual of Brazilian vegetation.** Rio de Janeiro: IBGE, 1991. 92 p.

. **Demographic census:** preliminary results - Natal. Rio de Janeiro, 2010. Available at: <http://censo2010.ibge.gov.br/>. Accessed on: June 28, 2012.

. **2010 demographic census:** Subnormal settlements - territorial information. Rio de Janeiro, 2010. Available

at:<ftp://ftp.ibge.gov.br/Censos/Censo_Demografico_2010/Aglomerados_subnor more/Aglomera

dos_subnormais_informacoes_territoriales/aglomerados_subnormais_informacoes_t erritoriales.pdf

>. Accessed on: 28 Nov. 2013.

National Institute of Meteorology. **Climatological Normals of Brazil 1961-1990**:

Revised and expanded version. INMET. 2010.

MARANDOLA, E. A geographical ontology of risks: two scales, three dimensions.

Geografia. Rio Claro, v.29, n. 3, p. 315 -338, Sep/Dec 2010.

MARANDOLA, E.; HOGAN, D. J. Natural hazards: the geographical study of risks and dangers.

Ambiente & Sociedade, Campinas, p.95-109, Jul/Dec 2005.

The dimensions of Vulnerability. **Sâo Paulo em Perspectiva**, Sao Paulo, v. 20, n. 1, p.33-43, jan/mar. 2006.

MEDEIROS, T. H. L. **Geomorphological evolution, (de)characterization and forms of use of the lagoons of the city of Natal-RN**. Master's dissertation. Natal-RN: UFRN, 2001.

NASCIMENTO, F. R. do; SAMPAIO, J. L. F. Geografia Fisica, geosistemas e estudos integrados da paisagem. **Revista da Casa da Geografia de Sobral**, Sobral, v. 6/7, n. 1, p.167- 179, 2004/2005.

NATAL (Municipality). Constitution (2007). Complementary Law No. 82, of June 21, 2007. **Natal Master Plan**. Natal, RN, 2009.

Natal. Acquatool consultoria. Prefeitura municipal do Natal (org.). **Plano Municipal de Reduçâo de Risco do Municipio de Natal**. Natal: SEMURB, 2008. 1,2 and 3 v.

Natal. L.R Engenharia e Consultoria LTDA. Natal City Hall (org.).

Drainage and Stormwater Management Master Plan - Natal/RN. Natal: SEMOPI, 2009.

OLIVEIRA, F. F. G. de. Characterization and diagnosis of environmental impacts in Natal/RN using geoprocessing. 2003. 207 f. Dissertation (Master's Degree) Geography Course, Department of Postgraduate Studies and Research in Geography, Federal University of Rio Grande do Norte, Natal, 2003.

UN (Rio de Janeiro). United Nations in Brazil. **UN:** more than 70% of the world's population will live in cities by 2050. 2013. Available at:

<http://www.onu.org.br/onu-mais-de-70-da-populacao-mundial-vivera-em-cities-by-2050/>. Accessed on: January 23, 2013.

PEREIRA, V. H. C. Areas vulnerable to flooding around the Neôpolis lagoon in Natal/RN: delimitation and mapping. IN: Symposium on Geodetic Sciences and Geoinformation Technologies, 4, 2012, Recife. **Proceedings of the IV SIMGEO.** Recife: CTG/UFPE, 2012. p. 1 - 5. CD- ROM.

PFALTZGRAFF, P. A. dos S. **Geodiversity of the state of Rio Grande do Norte** / Organized by Pedro Augusto dos Santos Pfaltzgraff [and] Fernanda Soares de Miranda

Torres. -- Recife: CPRM, 2010. 227 p.

NATAL CITY HALL. Public Cleaning: Solid Waste in the City of Natal. 1997. Available at: <http://www.natal.rn.gov.br/urbana/paginas/ctd- 128.html>. Accessed on: 02 Feb. 2014.

REBELO, F. **Geografia fisica e riscos naturais**. Coimbra: Coimbra University Press, 2010. 215 p.

ROBERTIER, J. Vulnerability conditions and risk representations in Latin-America: Framing the territorializing urban risk. **Global Environmental Change.** n. 22, p. 391-398, 2012.

RODRIGUES, Cleide. Geosystemic Theory and its contribution to geographic and environmental studies. **Revista do Departamento de Geografia**: USP, Sao Paulo, n. 14, p.69-77, 11 abr. 2001.

SEADE. **indice Paulista de Vulnerabilidade Social.** Spaces and Dimensions of Poverty in the Municipalities of the State of São Paulo. 2008.

available to me:

<http://www.iprsipvs.seade.gov.br/view/pdf/ipvs/metodologia.pdf>. Accessed on: January 10, 2013.

Geological Survey of Brazil - CPRM. Rainfall atlas of Brazil. CPRM. 2010. Available at: <http://www.cprm.gov.br>. Accessed on: October 3, 2013.

- CPRM. **NATAL SHEET SB-25-V-C-V**. Recife: CPRM, 2012. Scale: 1:100.000

Available at: <http://www.cprm.gov.br>. Accessed on: September 4, 2013.

SMITH, K. **Environmental Hazards:** Assessing Risk and Reducing Disaster.

London: Routledge. 2001.b

SILVA, A. F. C. da. LAND PARCELING AND THE FORMATION OF SPACES OF

POVERTY IN NATAL-RN.Scripta Nova, Barcelona, v. 7, n. 146130, 1 aug. 2003. Available

At: <http://www.ub.edu/geocrit/sn/sn-146(130).htm>.Accessed: Oct. 16, 2011.

THYWISSEN, K. Core terminology of disaster reduction: a comparative glossary. In:

BIRKMANN, J. **Measuring vulnerability to natural hazards.** Towards disaster resilient societies. Tokyo, New York, Paris: UNU-Press, 2006.

TRICART, J. **Ecodynamics.** Rio de Janeiro: IBGE, Technical Directorate, SUPREN, 1977. 91 p.

TRICART, J.; KILIAN, J. **Eco-geography and the planning of the natural environment**. Barcelona: Editorial Anagrama, 1979. 288 p.

VEYRET, Y. **The risks: man as aggressor and victim of the environment**. Sao Paulo: Contexto, 2007.

VIANELLO, R.L.; Alves, A.R.: **Basic Meteorology and Applications.** Viçosa - MG. University Press, 1991. 449p.

ZANIRATO, Silvia Helena et al. SENTENCES OF RISK: THEORETICAL INTERPRETATIONS. **Revista BibliogrAfica de Geografia y Ciencias Sociales,** Barcelona, v. n. 785, p.1-15, May 2008. Available at: <http://www.ub.edu/geocrit/b3w-785.htm>. Accessed on: January 25, 2013.

MIX
Papier aus verantwortungsvollen Quellen
Paper from responsible sources
FSC® C105338

Printed by Books on Demand GmbH, Norderstedt / Germany